Water, Environmental and Corporeal Traditions in Buddhism

This book examines ancient Buddhist traditions centred around water. It studies Buddhist monastics' sustainability and conservation practices and examines how these early practices and ethos apply in the contemporary world. The author draws on literary and archaeological sources across cultures and religions to trace the different socio-economic and cultural traditions that utilise water to formulate a pathway to save water in its purest form.

A unique contribution, this book will be an essential read for scholars and researchers of environmental studies, climate change, sustainable development, religious studies, and Buddhist history. It will also be useful to environmentalists and policymakers.

Anand Singh, Professor, Department of Historical Studies & Archaeology, Central University of South Bihar, Gaya, India.

Water, Environmental and Corporeal Traditions in Buddhism

Anand Singh

Routledge
Taylor & Francis Group
LONDON AND NEW YORK

First published 2025
by Routledge
4 Park Square, Milton Park, Abingdon, Oxon OX14 4RN

and by Routledge
605 Third Avenue, New York, NY 10158

Routledge is an imprint of the Taylor & Francis Group, an informa business

British Library Cataloguing-in-Publication Data
A catalogue record for this book is available from the British Library

ISBN: 978-1-032-76326-2 (hbk)
ISBN: 978-1-032-98896-2 (pbk)
ISBN: 978-1-003-60124-1 (ebk)

DOI: 10.4324/9781003601241

Typeset in Sabon
by SPi Technologies India Pvt Ltd (Straive)

Contents

Figures

Foreword

Water, the essence of all living beings and the Earth, is more than a physical element. Its cultural viability and social reality integrate social groups, shape the future of human civilisations, and form their very existence. The meanderings of water in majestic rivers and oceans, their crystals on mountains and glaciers, and blissful storage in lakes and ponds sustain the flora and fauna of the planet Earth. Humans, animals, reptiles, plants, and herbs seek their survival from water. Water has been a social and cultural milieu for humans since ancient times, but the era of industrialisation changed the notion of water. Now, it is recognised more in terms of material substance, rather than social and sacred. Due to a lack of proper vision, rising population, environmental decay, and consequent ecological crisis, the world is grappling with water problems. Millions of people still lack sufficient water to drink. The unsustainable use of water and its sources has disturbed the delicate equilibrium of the water cycle, and the world is facing unseasonal floods, droughts, cloud bursts, rising sea levels, and faster glacier melting.

The monograph addresses this complex issue by explaining the characteristics of water, environment, and Buddhism. The teachings of dependent origination, karma cycles, mindfulness, and compassion offer valued understandings into our relationship with water and the environment. It delves into the multifaceted linking between water, environment, and the teachings of the Buddha. It offers unwavering answers on water crisis and its solutions through a nuanced elucidation of Buddha's teachings, scientific research, and real-world experiences. I hope that my effort in the form of this monograph will sprout a new kind of thinking and stimulate meaningful conversations to ensure sustainable water resources for all.

During this academic journey, Professor Asanga Tilakaratne, Colombo University, always inspired me to carry out this work, I am thankful to him for his valuable advice and comments. Professor Manisha Chowdhry, Delhi University, and Dr Gyan Prakash, IIT Dhanbad, helped me complete this work in many ways. I thank both of them for providing me academic help from time to time. Dr. Srikant Ganvir, Professor Sushma Trivedi, Dr. Sudarshan Chakradhari, and Dr. Sharmistha Chatterji helped me in this scholarly journey. My sincere thanks go to each of them. Dr Stanjin Mingur and Mr. Nishant Kumar enabled me to obtain images and read the manuscripts. Blessings to both of them. Thanks to Archaeological Survey of India and Indian Museum Kolkata for getting epigraphic materials and pictures.

I sincerely thank Routledge publication for publishing this monograph.

Lucknow
15-03-2025
Anand Singh

1 Introduction

The beginning of organised life and the development of community, society, kinship, and language ushered in adherence by humans to the values of cultural ecology. They believed that natural forces were divine and that for the survival and sustenance of their cultures, propitiation and supplication of these natural forces were the greatest necessity. They failed to understand the mysteries of natural phenomena like thunder, rain, forests, clouds, water, rivers, oceans, etc. and began to worship them as mysterious and unknown spirits. The earliest form of worship viz., animism and animatism emphasised the importance of nature and its attributes. These were considered divine and different kinds of religious beliefs and rituals were developed to placate them. Because of such peculiarities and the need to coexist with them, humans developed different kinds of sacred spaces and consequently different religious systems, and through such sacred spaces, the human world made vigorous efforts to integrate and transform the unknown, mysterious, and mystical universe into the known ecosphere. These sacred spaces signify the ways and means to reach the unknown.[1] Water is the most important natural element that sustains the whole universe. Because of that people of the hoary past considered it divine and primordial. Different kinds of interpretations were made by sages and philosophers about the status of water. It has been considered static, flowing, wavy, regenerative, mystical, and mysterious. When it flows, it is like a life and perceived as 'live water' while its static form represents the mystical and regenerative attributes and is used for life support, rituals, and cleansing.[2] Water is perhaps the most sacred insignia of holiness, worldliness, and unworldliness. It pervades the earth and proliferates in all directions, meandering on different parts of the earth, hills, and forests subjugating the continents and penetrating beneath the earth to move to *patālaloka* where *nāgas* made their abodes full of treasures and riches. It suffuses the earth expanding its wings in hills, rivers, lakes, ponds, lakes, etc. A large quantity of water goes out and elopes with the sea. Despite that, the seas never get fattened from the daily intake of water from different rivers.[3] It is the most precious element for the survival of any kind of living being. The advent of *Homo sapiens* paved the way to designate water in terms of sacredness as well as its corporeal significance. The early habitations of the

DOI: 10.4324/9781003601241-1

humans were near sources of water and then human beings began to settle on the plain lands near the oxbow lakes and natural reservoirs; the banks of rivers were their first choice. The oxbow lakes came into existence due to the meandering nature of rivers, a peculiar phenomenon of the flood plains of northern India. These oxbow lakes and their fertile tracts were considered the lifelines of the early settlers across the world. The habitation in the vicinity of water gave men easy access to water for domestic use and irrigation. The monsoon rains, seasonal floods, and occasional droughts compelled them to ponder over the mystical and mysterious nature of water, consequently developing religious notions and means to placate or mechanise the water sources. The requirement of water for domestic use, irrigation, and trade led the architects of the earliest civilisations to develop and settle on the banks of rivers. The three earliest civilisations, that is Mesopotamian, Egyptian, and Harappan Civilisations, were born and developed on banks of rivers. The Mesopotamians and Sumerians were settled on the banks of Rivers Tigris and Euphrates, the Egyptian Civilisation flourished on the banks of River Nile, and the Harappan Civilisation developed along River Indus, its tributaries, and the River Sarasvatī. The early cultures of China were developed along the banks of River Huang-He (Yellow River). The people of these civilisations built efficient water management systems, dams and levees, water reservoirs, and canals for their domestic use and economic development. They also followed rituals, paid homages and reverences to placate the mysterious forces of water, and also developed special structures for such rituals as the Great Bath of the Harappan Civilisation.[4] In later Indic and other traditions, ascetics and scholars accepted the presence of four or five natural elements which created the whole universe, that is, fire, wind, water, earth, and sky. The Vedic sages speculated the possible explanations of the origin of rivers and springs. Buddhism provided the more profound reason for the genesis of water sources and rivers. Since the reformative phase, the 16th century onwards, the sacred notion of water started withering away. With the emergence of the modern industrial world, the population is growing enormously and water is used in a calamitous manner without thinking about its implications and consequences for future generations. The world has started feeling the reverse cycle of Nature where inundations, cloudbursts, and droughts become more frequent and such occurrences have been linked to global warming and demographic shifts caused by humans.

Water as Sacred in Early Traditions

Natural elements like fire, earth, and water were propitiated and worshipped by the people of the ancient civilisations. The mystic qualities of these elements protect the people from unseen dangers. Many types of water bodies like rain, lakes, rivers, wells, and oceans are symbolised and worshipped. Their natural characteristics are considered mysterious and worthy of worship. Several civilisations and cultures considered water a source of the genesis of all living and non-living beings and worshipped water as the most sacred element on earth. Mircea Eliade describes it as '*fons et origo*', that is, the source of all possible creation and existence. It has the primordial attributes that manifest all kinds of seeds of creation. It is formless and contributes to giving birth to all kinds of beings who after completing their life cycle again return to the water. To dip in the water is accepted to reminisce the root of life as water has regenerative power that designs the destinies of all living beings. At every step of life from birth to the funeral, water plays a pivotal role in the life of humans. During the great deluge, it engulfs the mortal world and with its regenerative powers, water recreates the world with new characteristics. Water with its power of creation, assimilation, and destruction maintains the cycle of momentum of the universe.[5] Because of the propensities of creation, destruction, and regenerative powers, the myth of the great deluge (*mahāpralaya*) and recreation of the world from the primordial attributes of water is hypothesised across religions. The great flood has emblematic value, signifying the processes of cleansing and renewal and is seen as a vehicle for rebirth. It shows a shift and recycling of the world before the great deluge and after that.[6] The early civilisations probably developed this concept because their people faced mammoth difficulties in surviving and sustaining the fury of hazardous geography, especially water. They were settled on the banks of rivers and had to face regular floods that led to the destruction of life and property. They placated water by developing strings of rituals and prayers and also made subtle arrangements to save their habitations.

The purpose of defence walls surrounding the Harappan cities is difficult to locate but scholars prefer to say that these walls were built to save the city from natural calamities like floods and storms. The fortalice or defence walls are known features of the Harappan cities as evident from Harappa, Mohenjo-daro, Lothal, Kalibangan, Banawali, Rakhigarhi, etc.[7] The archaeological remains inform that these walls were approximately 12–13 metres thick in cities like Harappa and Lothal and served the purpose of saving the cities from flood as the habitation settled at the floodplains of Rivers Indus and Sabarmati, respectively.[8] The huge parapet wall and citadel were built at Dholavira during different phases of its evolution and development. The occupation was stretched northwards and later on, a massive structure of a fortified middle town, north of the citadel was established. The lower town was developed east of the earlier structures within the city wall. The fortified town of Dholavira is different from other cities of the Harappan Civilisation. The middle and lower towns are surrounded by massive defensive walls and important areas of the city are also fortified. Such massive and meticulous circumvallation that is 13–18 metres thick makes Dholavira a unique fortified settlement.[9] The town was located on the banks of small ephemeral rivers, namely Mansar in the north and Manhar in the south. The wall thickness at Dholavira exceeds the military purpose and safeguards from weapons like slingshots, bows, arrows, etc. It suggests that the parapet was built with a specific purpose.[10] The bastions and gateways in the wall at Dholavira show the purpose of transmitting significant public warnings in case of any natural eventualities or calamities. The gateways served the purpose of facilitating greater accessibility for trade and commerce which was a thriving phenomenon in the region.[11] Dholavira, being very near to the sea could have been prone to cyclonic disturbances and oceanic catastrophes. In the mid-Holocene sea level, the vulnerability of this area to cyclonic disturbances is examined. The sea level was higher at 6,000–4,000 years BP.[12] It is contemporaneous with the timing of the Dholavira settlement. It is found that the citadel was developed much earlier than the city fortification and its wall thickness is more than that of the city wall. The terrain design in terms of elevation setup proposes that the citadel area is nearer to the coast and could be the first structure where any oceanic tragedy would come if it happened. The strategic coastal setting and elevation of the Dholavira settlement on an island suggest its suitability to be a promising port city with a possible susceptibility to any oceanic problems. The coastal geomorphology of the Kachh region shows that the interior part of the coastline has characteristics that feel the consequence of tsunami waves when they become mingled with high tides and become dangerous.[13] The massive walls of the Dholavira might have been for a protective arrangement against possible natural calamities like tsunamis/storms.[14] Similar kinds of mechanisms are known in other civilisations.

Because of its divine and primordial characteristics, water is considered sacred, and springs, ponds, lakes, rivers, and oceans are linked to deities and worshipped since the hoary past. To placate them, different religions developed several rituals, chants, and prayers and made offerings to them. The sacred cult of water or hydrolatry is popular among many cultures of the world. They are not uniform and are generally linked to the local traditions where they were developed. Because water was among the first needs, its worshipping was well embedded in their respective cultures.[15] The earliest civilisations were mostly agrarian, well supported by their industrial productions and trade networks. The people of the Harappan Civilisation harnessed plenty of water for domestic use and economy. They also developed a hydrological mechanism to collect the water in their great reservoirs and made the dockyard for the smooth sailings of their trading ships. The sacredness of water is well exhibited in the archaeological remains of the Great Bath found in Mohenjo-daro. It is approximately 12 metres long north–south, 7 metres wide, and 2.4 metres deep. Wide steps from two sides join the tank from the north and south and small openings at the edges of the steps might be to hold wooden planks. At the bottom of the stairs is a small ledge or bracket with a brick border. It covers the whole surroundings of the pool. The pilgrims coming down the steps could tread along this ledge without actually stepping into the pool itself. The floor of the tank is watertight due to impressively fitted bricks laid on the edge with gypsum plaster and the side walls were constructed similarly. The four walls of the pool were uniform and each had a thickness of 1.36 metres. To make the tank more watertight, a profuse layer of bitumen having a thickness of 2.4 centimetres was filled in the crevices of bricks, on the sides of the tank, and seemingly also beneath the floor. The structure is considered an epitome of ancient times.[16] Brick arcades are found on the eastern, northern, and southern edges. The well-preserved columns have stepped edges that may have wooden canopies or window frames. Two large doors lead into the complex from the south and the other entrances were from the north and east. The Great Bath may also have had one upper storey as evident from the remains of stairs, charcoal, and timber.[17] John Marshal says:

> *The Great Bath was a vast hydropathic establishment and the most imposing of all remains unearthed at Mohenjo-daro. Its plan is simple: an open quadrangle with verandahs on its four sides, and at the back of the three of the verandahs various galleries and rooms; on the south a long gallery with a small chamber in each corner, on the east a single range of small chambers, including one with the well; on the north a group of several halls and fair-sized rooms. In the centre of the open quadrangle is a large swimming bath some 39 feet long by 23 feet broad and sunk 8 feet below the paving of the court, with a flight of steps at either end and at the foot of each a low platform for the convenience of bathers who might have found the water too deep.*[18]

It is not known what kind of sacred beliefs and rituals the Harappan people developed for water but the structure of the Great Bath shows that they had a firm belief in the sanctity of water. The excavation of Inamgaon, a Chalcolithic site that dates back to 1000 BCE shows stone rubbles and mud embankment connected with the channels. It seems these water channels were used for religious as well as economic pursuits.[19] The communitarian nature of water is the utmost prerequisite for the survival of the whole animal kingdom. The utilitarian characteristics and life-sustaining features make water sacred and divine. The ancient civilisations practised bathing as a measure of cleanliness as well as hygiene. They built baths for the rituals and community bathing. When they came to take a dip in the community pool, though they came from different socio-economic strata they felt shared belongingness to a unified culture.[20] The sacredness and primordial characteristics of water are well explained in later traditions and rivers as an important source of water are worshipped. The *Ṛigveda* informs that all forms of life evolved from Apah, the water god and all of them merged with him. Because of that, he is considered the bestower of rain and the sweet essence of earth.[21] It also informs that the rays of the sun cause rain and the sun also extracts water from all parts of the universe. It begins creation through fire which continuously extracts and discharges water.[22] The tenth *maṅdala* of the *Ṛigveda* says that with the great churning the water originated first with the *Hiranyagarbha* and other species and objects were born.[23] In *Yajurveda*, water is the mother of the earth, and she revolves with water in her orbit. Sun is her father and he takes rounds of her.[24] In *Ṛigveda*, Sarasvatī is mentioned as a great river and provider of knowledge and insights to the sages for the creation of hymns. She is a divine river originating from the mountain and descending into the Indian Ocean crisscrossing a long distance. It flows fast with all sustaining great source of water and bestowing boons to the vast region.[25] It was supposed to be an earthly river that gave immense support to the *Ṛigvedic* Aryans. It was the main reason that the Vedic sages paid utmost reverence to this river. The *Ṛigveda* prays Sarasvatī as a river and a deity and she is invoked for food, riches, and progeny.[26] The tenth *maṅdala* of the *Ṛigveda*[27] indicates her dual character as a river and as a goddess along with other rivers like Sarayū and Sindhu. The *Ṛigveda* mentions that before the creation of the universe nothing existed except water. The *Nāsadiya Sūkta* of the *Ṛigveda* informs that all life seems to arise from the golden embryo, that is, the *Hiranyagarbha* or the Golden Womb which floated upon the surface of the primaeval waters.

Then even nothingness was not, nor existence,
There was no air then, nor the heavens beyond it.
What covered it? Where was it? In whose keeping
Was there then cosmic water, in depths unfathomed?

Then there was neither death nor immortality
Nor was there then the torch of night and day.
The One breathed windlessly and self-sustaining.
There was that One then, and there was no other.

At first, there was only darkness wrapped in darkness.
All this was only unillumined water.
That One which came to be, enclosed in nothing,
arose at last, born of the power of heat.[28]

The *Nadīstuti Sūkta* of the *Ṛigveda* eulogises the rivers Sindhu, Śatudru, Ganga, Yamuna, and Sarasvatī, among other rivers. It says:

Fair ye this my laud: O Ganga, Yamuna, O Sutudri, Parusni and Sarasvati:
With Asikni, Vitasta, O Marudvrdha, O Arjikiya with Susoma hear my call.
First with Trstama thou art eager to flow forth, with Rasa, and Susartu, and with Svetya here,
With Kubha; and with these, Sindhu and Mehatnu, thou seekest in thy course Krumu and Gomati.
Flashing and whitely gleaming in her mightiness, she moves along her ample volumes through the realms,
Most active of the active, Sindhu unrestrained, like to a dappled mare, beautiful, fair to see.[29]

The names of the rivers were added and more rivers were considered divine and sacred. These are cognizance of the personified forces of decontamination and regeneration. The *Śatapatha Brāhmana* says that water is the genesis of all existence, and it gives life, longevity, and healing.[30] Eliade says that when Hindu cosmology evolved then Viṣnu is represented lying on a lotus on the primordial water with the cosmic tree emerging from his navel. It symbolises the creation of the universe including humans and plants.[31] Ananda K. Coomaraswamy observes that aquatic cosmology/cosmogony became a popular subject in the Indic religions and its iconographical sketches are exuberantly represented in Brāhmanism, Buddhism, and Jainism where plants as a symbol of life and fertility are frequently depicted coming out from the mouth or navel of the *yakṣa*, *makara*, and the *pūrṅaghata*.[32] South Indian tradition informs that sage Agastya and Vaśiṣtha were born in the water pot as the union of Mitra and Varuna with *apsarā* Urvaśī. Because of that, he was known as *kumbhsambhava*, *kumbhayoni*, and *ghatodhbhara*. In other traditions, Agastya was the grandson of Brahmā and the son of Pulastya. He drank the water of the ocean and also married his daughter.[33] The *Chaṅdogya Upaniṣada* tells us that water pervades everything big or small, the earth, the atmosphere, the heaven, the mountains, the gods, men, animals, birds, grasses, plants, insects, ants, etc. All these are the worldly manifestations of water.[34] The cosmological nature of water is also well discussed in the later period. The movement of waves of the earth was reflected as *saṁsāra*; the roaring sound of water waves gives new life to vegetation. Water has healing properties that make humans healthy and free from diseases. In Indian religious traditions, rivers are worshipped for their regenerative powers, capabilities to remove sin, and to bequeath merits. In Buddhism and Jainism, the water of rivers is considered a boon for human beings, and their strength and purity are great aspirations in the course of meditation and attainment of enlightenment.

The sacredness of River Gaṅgā in Brahmanical tradition began with her descending on earth from heaven. The great Ikṣvākus made tremendous efforts to placate her to come to earth. The story began with King Sāgara who performed the Aśvamedha sacrifice and whose *yajna* horse was led by his grandson Ansumant. As an eternal part of Indian mythological tradition, Indra played a mischievous role in mitigating the growing divine influence of King Sāgara. He stole the sacrificial horse and hid it in the nether world. The 60,000 sons of King Sāgara started searching for the *yajna* horse in every nook and corner of the earth, ocean, and sky. Due to the fear of the wrath of Sāgara's sons, Indra left the horse in the *āṣrama* of sage Kapila. When Sāgara's sons discovered the horse, they became enraged presuming that the horse was caught by Kapila. The sage without knowing the reality reacted angrily and burnt them down. When Ansumant informed the incident to his grandfather, he felt deep sorrow for losing his sons. He was informed that only the sacred water of the Gaṅgā could liberate the souls of his 60,000 sons.[35] Ansumant performed great austerities but failed to bring

Gaṅgā to earth. His son Dileep, another great Ikṣvāku scion performed great austerities on the Himalayas for 3200 years but could not win the favour. After his death, his son Bhagiratha decided to placate Gaṅgā on earth to pay libation for his 60,000 ancestors. He went to the Gokarna hill and performed austerities and won favour from God Brahmā who advised Bhagiratha to worship Śiva to hold the mighty force of Gaṅgā. Bhagiratha prayed to Śiva who consented to hold the snow-clad river in his matted hair locks. In the company of Lord Śiva, Gaṅgā became full of life-giving attributes and regenerative powers. A dip in the holy waters of Gaṅgā can remove all kinds of sins and specks of dirt. When Gaṅgā descended on the earth, Bhagiratha led the way on his chariot accompanied by the company of gods, sages, and nāgarajas. Gaṅgā washed the ashes of the 60,000 sons of the Ikṣvākus and all of them went to heaven. With the blessings of God Brahmā, Gaṅgā became identified as Bhagirathi, the daughter of Bhagiratha. She traversed 500 miles, reached Haridwar, the abode of seven sages (*saptarishi maṅdala*), and then moved in seven different directions. When she unknowingly drenched the *āṣrama* of sage Jahānu, he became enraged and drank all the water. Bhagiratha again rescued her and took all the water from Jahānu's knee. After that, she became Jahānvī. Gaṅgā meandered through the northern plains of modern Uttar Pradesh and Bihar and finally traversing the different paths in the Bengal delta, she merged with the ocean.[36] Gaṅgā became a mother, a bestower of fertility, and prosperity across India. Though her water irrigates the northern land, she is worshipped in all parts of India. The most fascinating visual tradition of descending River Gaṅgā is not found in north India but in the temple of the Pallavas in the south Indian plain irrigated by the monsoon River Pallār. The Pallava kings Mahendravarman I, Narsinghavarman, and others made Mamallapuram (Mahābalipuram) and Kanchipuram great centres of art. The temple of Mamallapuram shows a relief of River Gaṅgā descending from heaven through the matted hairlocks of Śiva. She is flanked on either side by gods, sages, and life-size elephants. After a gap, the anthropomorphic figures of nāgarājas are sculptured.[37] The images of elephants and nāgarājas are considered symbols of prosperity, rain, water, and fertility.

The Kumbh Melas, held over a cycle of 12 years, are hailed as the largest congregations of humans for baths. These congregations convened on regular intermissions at Haridwar, Prayāga, Ujjain, and Nasik along the banks of River Gaṅgā; the confluence of Gaṅgā, Yamunā, and Sarasvatī; Kṣipra, and Godavarī respectively. The tradition seeks nomenclature from the divine pitcher (*kumbha*) which was filled with *amrita* or nectar. Due to the spat between the Devas and *Rakṣas*, some parts of it were spilt in different places. These four places are the most important. Taking a dip in the holy waters of these sacred rivers during the Kuṁbh Melā is considered auspicious in Hindu tradition. River Gaṅgā is specially worshipped at the Gaṅgā Sāgara festival held every year in West Bengal, at the confluence of River Gaṅgā or Huglī and the Bay of Bengal (popularly known as Gaṅgāsāgar). Millions of pilgrims visit every year from all parts of the country to take a bath.[38] Gaṅgā as a mother and a nourisher became a synonym for many rivers in India. Gaṅgā is like the paradigmatic sacred river which is not restricted to the path she traverses across the plains of north India but also carries her presence in the vast Hindu sacred topography, permeating the sacred waters of the great rivers of India.[39] Her influence could be seen not only in India but also in Southeast and East Asia. River Mekong is also portrayed as 'mother Ganga' by people of Southeast Asia. It traverses the path covering the vast topography of China, an autonomous region of Tibet, Myanmar, Laos, Thailand, Cambodia, and Vietnam.[40] Mekong is the main source of Buddhist culture which is permeating the life and livelihood of people living in the area where she is meandering. In Khmer tradition, 'Me' means mother and 'Khongk' is derived from Gaṅgā. Because of that, River Mekong is known as Mother Gaṅgā. It is the source of inspiration, livelihood, and spirituality for more than 66 million people living in that region. It is still flourishing in lavishly lush forests, mist-covered hills, and the music of water flow in the Mekong. Buddhism is a predominant culture of the countries situated in the region dominated by the river. The folk traditions and religious narratives of the region show that the river was born due to the sacred power of the Buddha. Originally Mekong was a large lake surrounded by hills but the Buddha facilitated the way to flow the water in the plains of Southeast Asia.[41] Another tradition informs that the Buddha used the miraculous sword to cut the hill ranges obstructing the flow of the river. The sacred streams coming from different hills joined together and formed River Mekong. The nāgas and catfish of the region helped to draw channels from the river[42] Analogy is drawn between Rivers Gaṅgā and Mekong. The different streams coming from the Himalayas merged and formed River Gaṅgā; similarly, the different streams from the same mountain formed River Mekong. The only difference is the region and directions where these rivers are flowing. The Buddha like King Bhagiratha put the same effort into taking River Mekong to the plains. Though he did not perform rigorous penances, his *iddhi* played a decisive role. Across the countries around River Mekong, the *Buddhapadas* are still the main source of worship and people still believe that the Buddha himself traversed the path when River Mekong originated and the remains of the footprints belong to him.

The depiction of Gaṅgā was not as a river in religious traditions but as a sacred force that creates and sustains the people of India with her divine regenerative powers. In Buddhism, River Gaṅgā is equally eulogised but not as ridder of sins. She is linked with the creation of this *saṁsāra* and our great civilisation. Gaṅgā along with Rivers Yamunā, Acirāvatī, Sarabhu, and Māhī are the five great rivers that originated from the divine lake Anotatta in the great Himalayan region.[43] Here, the origin of Gaṅgā is less mythical and her attributes are the best gift for the *saṁsāra* and also to the monks for meditation and attainment of *jhāna*. In the Indian religious context, the corporeality of rivers is widely incorporated, embedded, and mystified with different kinds of emblematic, metaphysical, epistemological, and transcendental linkages.

Rivers like Gaṅgā, Yamunā, Sarasvatī, Godāvarī, Kāverī, Narmadā, Saryū, and others are personified as deities, and various cultic traditions are developed on their name. River Godāvarī is known as Dakṣina Gaṅgā and Kāverī also earns the epithet of Gaṅgā of the South. Both are rainfed rivers and originate in the peninsular region of India.[44] Both carry rich heritage and culture across South India.

The people of non-Indic cultures and civilisations also believed in the sacredness and primordial characteristics of water. The people of Babylonia also believed in the primordial characteristics of water and considered that the primordial ocean *Apsu* is made of freshwater and that the earth was made to float on it. The other was *Tiamat*, a salty and bitter water sphere that was occupied by monsters.[45] They believed that all the lands were part of the sea. Marduk, the legendary Babylonian creator covered the face of water and facilitated its accumulation.[46] The Egyptians also accept that the water on earth is perennial, and the universe was created because of the primordial water of Nūn.[47] The early Greeks did not show much interest in the cosmological attributes of water. Their geographers thought that the planet Earth was surrounded by a mighty stream Oceanus. Homer mentions it in the *Odyssey* and the *Iliad*. According to this belief, water is well separated by sea and its genesis is mysterious. Oceanus is the source of origin of great stars except for the constellation of the Great Bear. Herodotus informs that locals believed that River Nile originated from Oceanus. Titan Oceanus, the son of Uranus (heaven) and Gaea (earth) was one of the natural forces responsible for the creation of the universe. Homer places him among the pantheon of the gods, only second to Zeus. He married his sister Tethys and had progenies – 3,000 rivers and 3,000 Oceanids (sea-nymphs). Because of that, he is worshipped as the father of all the rivers, seas, and other kinds of primaeval waters.[48] The Greek culture was developed under the prodigious influence of Ionians in Asia Minor but Egyptian influence is also quite visible. Thales of Miletus was the first who deliberated on the water cosmogony. He believed that earth floats on the water. It is the first or original substance (*arché*) and because of that everything comes out of it.[49] The Greek philosophers kept searching for the first element which was born and then the universe was created. Heraclitus presumed that fire is the first element and later on, this idea was broadened by ritualists who included water and believed in the dyad of water and fire as *archés* or primary substances. Euripides believed that air and earth were the primary substances. Empedocles of Agrigentum believed in the four elements (*rhizomatā*) – fire, air, water, and earth. Asit Biswas infers that the idea of four elements among the Greeks comes from Pythagoras.[50] Anaximander of Miletus (610–545 BCE) also believed in the sacredness of water and that all living beings originated from water. During the process of evaporation, the earth emerged out of the sea. He also gives the reasons for rain, the making of snow, and hailstorms. The cause of rain is moisture absorbed from the planet Earth by the Sun. The hailstorms come when the rain becomes frozen and ice is formed when air is compressed within the water.[51] Xenophanes (528 BCE) believes that water and wind originated from the sea. Besides that, the source of origin of clouds rains. Spring and streams are water. [52] From the 5th century BCE onwards, Socratic views dominated the arena of Greek sciences and philosophy, and the ages of Plato and Aristotle followed them. Plato accepts the genesis of the universe with four elements fire, air, water, and earth which was well accepted by his predecessor Empedocles of Agrigentum. Empedocles borrowed the concept of four *archés* from Assyria. Plato added a fifth element which he called heaven.[53] It is also deplorable whether the theory of the four elements fire, water, air, and earth was developed in Assyria or not. In the Indian subcontinent, *catudhatu* or *caryamahābhūtas* were well accepted. Buddhism accepts that the universe is created by fire (*Tejo*), air (*vayo*), water (*apo*), and earth (*pathvi*). These elements are the foundation for the understanding of suffering and its liberation. The Pali texts mention that the four primary material elements are the sensory qualities solidity, fluidity, temperature, and mobility and their characterisations are earth, water, fire, and air, respectively. The Buddha's teaching regarding the four elements is to be understood as the foundation of annotations of real sensations rather than as metaphysical assumptions. The four characteristics are cohesion (water), solidity or inertia (earth), growth or vibration (air), and heat or latent energy (fire).[54] The Buddhist notion of the four elements predates the Greek teaching of the four elements, that is *mahābhūtas* or *archés*. Plato proposes that the great ocean of Tartarus lies in the subterranean region. It is considered to be the largest reservoir that circumferences the entire earth. The pool of its water has neither a bed nor a bottom and its waves always move up and down. The seas, rivers, lakes, and springs are born from the subterranean channels coming out of this great ocean and the water of these sources goes back to the source of origin.[55] Aristotle (385 BCE) also believes in the five elements. According to him, Earth was in the centre of the universe and surrounded by water. He believes that rains come from the condensation of a substantial amount of vapour for a long duration over a huge area and dew is also produced by a small amount of vapour collected in a single day. Only the lighter and sweeter part of the water was evaporated by the Sun. The freezing of clouds forms snow. He borrowed Plato's idea that seas, rivers, lakes, and springs originated from a great subterranean reservoir but differed in the size of the reservoir.[56] Such beliefs and practices proliferated in different modes like worshipping springs, rivers, ponds, and wells when the Greek and Roman civilisations expanded. Many of such practices in their modified forms and outlooks are still practiced in different parts of the world.

Water as Corporeal: Contexts in Early Traditions

Scientifically, water is an amalgamation of two atoms of hydrogen and one atom of oxygen. It is transparent, without any colour, smell, or taste. Water lies between ice and vapour and continuously changes its form depending on the season, climate, and temperature. Seasonal variations, physical parameters, and natural changes transform its status like temperature converts water from liquid to vapour and the lowering of temperature alters it into ice. Its modern structure and nomenclature H_2O were done by French scientist Antoine Lavoisier in 1783. It has different forms in different conditions and can be seen in drops of rain, rivulets, ponds, lakes, oceans, tears, sweat, saliva, urine, serum, etc.[57] It is considered a liquid that comes from the cloud in the form of rain, making rivulets, lakes, rivers, and oceans. The underground water comes out from springs. Freshwater is a valuable source for human survival. Its physical feature consists of an oxide of hydrogen with two atoms of hydrogen and one atom of oxygen. It freezes at 0^0 and boils at 100^0 and its maximum density is 4^0.[58] Jamie Linton argues that the discovery of the physical formation and molecular structure of water changed its sanctity and belief that has continued since the ancient past. It was always H_2O but its new interpretation in the scientific domain broke the sacred beliefs about water carried by the different traditions. Now water is treated as a physical entity in terms of hydrological and economic terms. It is considered an unlimited source that can be used by more than 8 billion humans, animals, and plants across the world. In ancient and medieval societies, there were different kinds of rituals, patterns of usage, cosmic bonds, and prayers for water. It depends on the availability of water in a particular region. Where water was abundant, people prayed for timely rain, avoidance of floods, and local mechanisms of making and preserving ponds and lakes to conserve water for summers. In case of scarcity of rain and water, people worshipped for abundance of rain and invented special kinds of reservoirs for rain harvesting.

The wells as a source of water are considered holy across the globe. In India, different kinds of rituals are performed on wells on the occasion of birth, marriage, and death. In the Indian subcontinent, wells were the most important source of water for domestic use. In the Harappan Civilisation, an abundance of wells has been discovered in cities like Mohenjo-daro, Harappa, Rakhigarhi, etc. In Mohenjo-daro, almost every household has the provision of a well and nearly 700 wells have been discovered from here. These wells are lined to avoid caves in and to get clean and sweet water without any silt and other particles.[59] The wells had platforms made of bricks or brick surfaces, used for bathing, cleaning, or rituals. Mainly four characteristics of a well are known – (i) without any brick or stone lining, (ii) with the lining of bricks, (iii) with rubble masonry, and (iv) with terracotta rings. Wells without linings are considered to be the first human effort to fetch water but the earliest archaeological remains of dug wells are represented by brick-lined circular wells dating back to the Harappan Civilisation. Ring wells with linings of terracotta rings and wells with linings of bricks had widespread distribution in space as well as in time. These still exist in different parts of India.[60] Some houses have multiple wells viz., in the courtyard, the bathroom, and the third one inside the kitchen. The practice of digging wells inside kitchens or houses is well narrated in Buddhist literature. One such episode from the *Mahājanaka Jātaka* is depicted in Ajantā showing a huge pot consisting of two rings fitted together.[61] The literary records also suggest that proper maintenance and care were taken to maintain hygiene and keep the water of the well clean. The *Vrhatsamhitā* informs that because of impurities the water of the well might become sour and distasteful and recommends its proper treatment. For purification and cleansing, a mixture of herbs should be put in the well. The mixture of *anjanam* (collyrium or antimony), *nagarmotha* (musta tubers), *khas* (*usira*), powder of *rajkostaka* (*torayi*), and *āmalaka* (myrobalan) with *kataka* nuts must be put into the well. The pungent odour, saltish taste, and other dirt will be wiped out.[62] In the semi-arid and desert regions of Persia and Armenia, a peculiar mechanism was developed to conserve and use groundwater. Artificial underground channels known as *qanāts* were built to channel the water over a long distance either from a spring or from other sources of water. Water was limited in the arid area and underground channels saved them from unnecessary evaporation. *Qanāts* also keep water cool and save from outside pollution.[63] In the West, wells had the same purpose. They served the water supply for domestic use and their water was also used to cure diseases and ward off the negative influences of witches and fairies.[64] In Christianity, the wells are considered holy and linked to local Christian saints or devoted to members of the holy family or St John. It is believed that the majority of the holy wells had pre-Christian origins. The references are found that saints took baths in the well at night, sang psalms, and rested on a nearby stone slab during the daytime. The key religious festival at the wells happened on midsummer when a large congregation of pilgrims took a bath naked in these holy wells. It has been argued that such religious practices were a continuation of pre-Christian midsummer rituals which highlight the therapeutic and regenerative powers of water. It is believed that immersion in the holy wells could lead to good health for humans and animals.[65] Rattue on the sources of literary and archaeological evidence suggests that well-bathing rituals existed in European cultures even in the period of the early Iron Age.[66] Some of the wells were developed from pagan practices and the Church intentionally imbibed the traditions of pagan sacred sites. Pope Gregory's letter to St Mettitus and fellow missionary St Augustine in 601 CE declares that the temples of the Pagan tradition should not be destroyed but only their idols must be. These shrines should be sanctified with the sprinkling of holy water, building of altars, and placing relics in them.[67] It suggests that pagan wells were used for rituals related to Christianity.

Canals, dams, and reservoirs are also important means to conserve water. One of the earliest examples of making reservoirs has been found in Dholavira in the Harappan Civilisation in the arid region of the Rann of Kachh. It is located on the banks of small ephemeral channels, namely Mansar in the north and Manhar in the south. Even though there are several other much deeper channels in the area, with comparatively large water-carrying capacity, the choice of the much smaller Mansar and Manhar for the settlement, points towards their suitability for water harvesting. These streams bear evidence of damming in the form of huge rock-cut pieces and boulders belonging to different geological formations that are too heavy to be transported by floodwaters. The construction of dams on other bigger channels would have been relatively difficult. The water-harvesting expertise of the people of Dholavira is also well exhibited in the form of reservoir systems, deep-dug wells, storage tanks, and storm channels built in and around the citadel. The extensive water storage structures were supposed to be built because of the limited seasonal availability of water, and therefore suggest a lack of perennial freshwater sources around Dholavira. R.S. Bisht says that it was a well-thought-out decision to build the city on the banks of such a small channel, as it can be dammed. Therefore, it is highly unlikely that the Mansar and Manhar channels posed any flood-related risk to Dholavira. The elevation of the Dholavira settlement is sufficiently high to avoid flood disasters.[68] The large reservoir with a stormwater drain is situated in the south and a series of reservoirs have been discovered and excavated in the south, west, and north of the habitation area. A feeding drain was made from the south of the reservoir. The shape of the largest reservoir is rectangular measuring 73 metres north–south and 30 metres east–west. The reservoir was provided with 30 steps each. The rainfall was scanty in Dholavira during the Harappan period and the only source of water was the preservation of rain-fed water and water from the two small seasonal water streams of the Mansar and Manhar. Dholavira had huge reservoirs, three of which are currently excavated and were supposed to be used for storing water fetched by rains or the water diverted from two nearby rivulets. It was based on sophisticated water management consisting of a network of water channels and reservoirs, built completely of stone. Their efficient mechanism for the conservation, harvesting, and storage of water expresses eloquently their progressive hydraulic engineering. The inhabitants of the Dholavira knew that depending on rainwater was not a dependable option to survive and sustain. They magnificently built large rock-cut reservoirs, located on the eastern and southern sides with large drains which directed the water to the western and northern sections of the lower town separated by broad bunds. That mechanism reflects a series of large water tanks. An imposing well is located in the fortress that might be the earliest example of a rock-cut well. Dholavira's drainage system was intricate, with many small and large drains intersecting each other to join the arterial drain.[69] During the second urbanisation and state formation, the making of reservoirs and digging of canals became sporadic in parts of the Indian subcontinent.

The earliest reservoir built for Buddhist monastics in the land of Magadha is found in the Giriyaka Hilltop monastery, Rajgir. An oblong water reservoir is found on the hill at the receding end of the higher slope within 50 feet of the main *stūpa*. The tank was excavated from two sides of the natural rock of the hill and the other two sides were plucked up with stone blocks. It is about 100 square feet.[70] There is a second tank at a short distance to the north formed by the excavation of the rock. One natural pond was at the bottom of the hill where the parapet had the foundation. The basic purpose of this tank was to harvest rainwater for the routine work of the monastery. Making reservoirs for rain harvesting was an important feature of the Buddhist monastic settlement. This structure was one of the earliest experiments in the land of Magadha. This technique was experimented with more broadly in the monasteries situated on the mainland of Magadha and an *ahara-paine* system of irrigation (based on rain harvesting) was developed. For monasteries on the hills, it was difficult to get water; so all such monastic establishments had the provision of making cisterns and reservoirs. The masonry and use of stone blocks suggest that its excavation was pre-Mauryan and definite beneficence by state patronage.[71] Probably, it was built by King Ajātśatru. In later phases, water reservoirs became an important feature in Buddhist and Hindu religious structures. The Hathigumpha inscription of Khārvela informs that King Nanda of Magadha dug a canal in Kalinga. The period of King Nanda was 4th century BCE.[72] The Hathīgumphā inscription also informs that King Kharvela also repaired the embankments of the lake later known as Khibira Rishi Lake and also caused the building of many tanks and cisterns.[73] Rudradaman's Junāgarh/Girnār inscription gives one of the longest histories of the construction and maintenance of water reservoirs for irrigation. The Sudarśan Lake and Dam were first built in the period of Chandragupta Maurya by his minister Puṣyagupta. In the period of Aśoka, his Yavana governor Tusaspa excavated channels for the benefit of the people. In the period of Rudradāmana, the dam of the reservoir was damaged and it was repaired by his minister Suviśākha on the instruction of the king, and for this maintenance, no tax was levied from the people.[74] The Girnar Inscription of Skandagupta is inscribed on the same rock and mentions that the lake was again repaired when the dam was broken due to a flood in River Palasinī. His minister Chakrapālita did commendable work for public welfare. The inscription says:

> *When, in the course of time, there came the season of clouds, tearing asunder with (its) clouds the season of heat, much water rained down unceasingly for a long time—in consequence of which (the lake) Sudarśana burst from the four roads.*
>
> *And now when a century of years, increased by thirty-six (had elapsed), making the calculation according to the Gupta era, at night, on the sixth day of (the month) Prauṣṭhapada— these (rivers) sprung from (the mountain) Raivataka, and this Palāśinī that moves gracefully in the sand, (all of them) the wives of the ocean, having dwelt for long in bondage, went again to their husband (the ocean), in due accordance with the scriptures.*
>
> *Noticing the great bewilderment caused by the advent of rains, (the mountain) Ūrjayat, wishing to do a good turn to the great ocean, stretched forth, as it were, a hand, consisting of the river (palāśinī), decorated with the numerous powers that grew on the edges of (its) banks.*

Feeling, indeed, dejected on all sides, discussing how they should act, keeping awake the whole night by turns, the people reflected (thus), feeling restless "just in a moment, (the lake) Sudarśana, has become disagreeable to the sight of men—can it ever become pleasing of aspect, having the appearance of the ocean....?

He, having become...., displaying exceeding devotion to (his) father, (and) placing, (prominently) before him, religion, (his) sacred ties to the king, and the well-being of the town;

When a century of years, increased by thirty-seven (had elapsed), comprehending the sacred writings,....whose great might is well known—though (he is) viśva;

Then having sacrificed to the gods with oblations of clarified butter and with obeisance's—and having gratified the twice-born with (presents of) riches—and having honoured the citizens with such honours as they deserved and (his) respectable servants and friends with presents;

In the first fortnight of the month (called Āṣāḍha) and belonging to the hot season, on the first day, he, having put forth careful efforts and made an immeasurable expenditure of wealth, in two months, laboriously built up a hundred cubits in all in-depth, and sixty-eight in breadth, and seven (?) men's height in elevation (of the breach into the embankment) of two hundred cubits, having done honour to the kings, built with great labour, with stone well laid, (so that) the lake, not evil by nature, became renowned as Sudarśana (of good appearance) for all eternity.

Agitated by the lower part of the body of the ruddy-geese, the herons and the swans which have displayed their beauty on the edges of the firmly built embankment....pure waters....(so long as) the sun and the moon.

And may the city become affluent—teeming with citizens—bereft of sin through prayers sung by many hundreds of Brāhmaṇas, (and free) for a hundred years from distress such as those caused by calamities (like) famine) Thus ends this literary composition of the restoration of the Sudarśana.[75]

Julia Shaw examined the Sānchī Buddhist Complex and observed that the community of monks and nuns aligned with local folk traditions. The local agricultural deities like the *nāgas* were appropriated in the Buddhist monastic tradition. It is widely believed that they control the monsoon which was an important constituent of Indian agriculture. With the blessings of the *nāgas*, they developed enormous facilities for irrigation by conserving seasonal rain and making reservoirs and dams.[76] From the 6th to the 7th century CE onwards, the Buddhist communities of Ladakh developed a unique irrigation system. Ladakh, *pod yul*, or *khong yul* (land of snow) is situated between the Indus and the Syok Rivers. It derives its nomenclature from *darras* (passes) comprising two words *la* (pass) and *dakh* (a range of hills). It extended from the Syok–Indus confluence up to the western border of Tibet where the Indus separates it from the Zanskar range in Pupshu. The Buddhist communities of Ladakh developed a highly efficient and ecological irrigation system. The people of Ladakh built canals with locally available materials, stone pieces, and boulders. The accumulated snow started melting as *kang-chhu* (ice water) in the rivulets merging at some focal points and forming a *togpo* (stream). The *togpo* flows in the valley touching the settlement clusters and the villages and is connected by the *ma-yur* (mother channel).[77] Numerous such examples are known in India when kings, officials, traders, and lay people dug wells and built canals for monasteries, temples, and irrigation.

The people of neighbouring civilisations were conscious of the preciousness of water. Chinese Emperor Yau, in 2280 BCE, instructed Yu to plan and build the big dams and dikes in China. He minutely observed and learned the viable mechanism for water conservation. He is considered a master of water and is instrumental in constructing dams and ponds. He was so popular in China that he was appointed the next emperor of China.[78] In Egypt, a drawing found on the mace head of King Scorpion shows the king inaugurating the digging of a canal and he was digging an irrigation ditch with a large hoe. In front of the king, two men are shown having a basket probably filled with seeds and a pile of corn, respectively. The king ruled in 3200 BCE. King Menes built his capital at Memphis in 3000 BCE and built a dam on River Nile at Kosheish to divert the course of the river to a newly built canal between two hills.[79] The archaeological remains of the dam Sadd el-Kafara (Dam of the Pagans) constructed between 2950 BCE and 2750 BCE near the south of Cairo, Egypt give some technical and architectural details of the water conservation technology of ancient times. It is presumed that this dam was raised to supply drinking water for the artisans and animals engaged in the alabaster quarry situated near it.[80] It was found that the reservoir of the dam has limited capacity to store the water and in the first heavy rain the water might have been overfilled. Its construction was vulnerable as no mortar was used and the lack of provision of a spillway made the situation difficult in case of overflow.[81] However, raising such a huge structure was a strenuous effort. King Amenemhet III or Moeris built a large canal between 1850 and 1800 BCE to manage the flood of River Nile. Its lake is said to have a circumference of 450 miles.[82] The Mesopotamians also developed mechanisms to control the floods and conserve and channel the water of the Tigris and Euphrates. The remains of the canal have been discovered by archaeologists.[83] The Code of Hammurabi, a compendium of Sumerian Law was discovered by French archaeologist Jean-Vincent Scheil in 1901. It also inscribes the code for the construction and preservation of dams and the conservation of water. In case of breaking of dams or overflow of water from ditches used for irrigation, a suitable penalty was levied.[84] The Marib Dam was considered one of the wonders of ancient Persia. It was situated in Wadi Dhana and was built between 1000 BCE and 700 BCE near the ancient city of Marib. It was 33 feet high and 1,900 feet long. Mortar was used to cover the top to save damage from excessive rain. It collapsed in the sixth century.[85] The Romans made aqueducts for the supply of water to their inhabitants. Earlier they were using water for domestic purposes from River Tiber, wells, or springs. Appius Claudius Crassus built the first aqueduct, the Appia in 312 BCE. The remains of architecture show the large arrangement of sluices, gates, long canals, and decorative motifs. In the distance, there were large size baths.[86] These dams, reservoirs, and baths played a decisive role in boosting the socio-economic fabric of states as well as the people. Big dams and canals boosted agriculture and the local economy. The wells and baths served the purpose of domestic requirements.

Karl Wittfogel studied agrarian societies and their irrigation pattern. Based on their irrigation pattern, he differentiates between small-scale and large-scale farming communities. The first group is engaged in small-scale self-sufficient irrigation systems, that is hydro-agriculture, and the other depends upon a large irrigation and flood control system, generally managed by the state. It is known as hydraulic agriculture. He says that the term 'hydraulic civilization' or 'hydraulic society' is more appropriate to represent the agrarian societies that depend on human-induced irrigation mechanisms rather than geography. Its systematic study discovers man's continuous efforts to get a suitable irrigation pattern in an arid, semi-arid, and humid climate.[87]

> *Water is heavier than most plants. It can nevertheless be much more conveniently managed. Unhampered by the cohesiveness of solid matter and following the law of gravity, water flows automatically to the lenient accessible point in its environment. Within a given agricultural landscape, water is the natural variable par excellence. And this is not all flowing automatically, water appears unending in the landscape, gathering either below the surface as groundwater, or above the surface in separate cavities (holes, ponds, lakes), or continuous beds (streams, rivers). Such formations are of minor significance in an agricultural area enjoying ample precipitation but they become immensely important in the water deficient land-scape.*[88]

The ancient structures built for the preservation and conservation of water for domestic and agrarian requirements must be examined holistically to understand their points of view. They were more ecologically oriented than the modern hydrological engineering developed to mechanise and monopolise water. The remains of water sources from pre-historical and historical societies suggest that they made only minor adjustments to nature and local conditions, contrary to the modern large-scale hydraulic projects of unbending models.[89] The cultural values, beliefs, and matrices of water in different cultures vary from one another. If one tries to superimpose one's belief over others, it will jeopardise the cultural equilibrium. A social definition of water cannot be universalised.[90] Vasudha Narayanan says that in the Hindu tradition, bathing in rivers on different auspicious occasions removes sins and replenishes the *puṅya* as rivers are considered mothers who not only give birth but also nourish. Gaṅgā is represented as the mother and the consort of Śiva, coming out from the great Himalayas and giving her blessing and fertility to the whole of north and east India. In Tamil Nadu, Kāverī is accepted as the consort of Viṣnu and after the coming of the new monsoon, the people of Śri Rangam believe that she is pregnant and celebrate her life-giving virtues. The meandering river with abundant monsoon water sweeps the agrarian plain of Thanjavur delta and waters the extensive paddy fields. The rivers in part of India have been considered natural purifiers and nourishers like mothers but rivers in this age are full of filth and industrial and toxic wastes. Because of that the waters of the rivers are losing their regenerative attributes and naturalness.[91] She tells:

> *The ashes of the cremated body are immersed in holy waters- the same rivers that feed and irrigate paddy fields; the same water that cooks the rice and baths the dead before cremation. From cradle to cremation, Hindus have long had a palpable, organic connection with nature. But today they also face the reality of environmental disaster with the population hovering around a billion in India, the use abuse and misuse of resources in placing India on the fast track of disaster.*[92]

It is important to sustain the sacred notion of water as people will understand its importance and its sanctity will be preserved for the sustenance of the future. Water is universal and shared by people and other species since time immemorial. Seeing water in material terms is disadvantageous to human society. The mythologies, folklores, traditions, and sacraments linked to water worship and rivers are intertwined into the syncretic and amalgamative cultural miscellany. The emergence of the modern industrial world, unification of the world economy, and the resultant crises like acid rain, melting of glaciers, drying up of rivers, and environmental catastrophes challenge the traditional sacred discourse linked to water. It is now important to examine the beliefs and traditions of water worship in its canonical as well as corporeal framework to understand and examine its tangible and intangible characteristics. It will help to find the cause for its perpetuation and the path of an interface between religious discourses about the sacredness of water and environmental virtues. These findings will be conducive to formulating and implementing the future discourses necessary to save water in its purest form.

Notes

1 Chris C. Park, *Sacred Worlds: An Introduction to Geography and Religion*, London: Routledge, 1994, pp. 246–247.
2 Ken Dowden, *European Paganism: The Realities of Cult from Antiquity to the Middle Ages*, New York: Routledge, 2002, pp. 7–8.
3 Asit K. Biswas, *History of Hydrology*, Amsterdam: North Holland Publishing, 1970, pp. 140–141.

4 Biswas, *History of Hydrology*, p. 1.
5 Mircea Eliade, *Patterns in Comparative Religion*, London and New York: Sheed and Ward, 1958, pp. 188–189, 210.
6 Nick Middleton, *Rivers: A Very Short Introduction*, Oxford: Oxford University Press, 2012, p. 34.
7 R. Nigam, R. Dubey, R. Saraswat, Sundaresh, A. S. Gaur and V. J. Loveson, 'Ancient Indians (Harappan Settlement) Were Aware of Tsunami/Storm Protection Measures: A New Interpretation of Thick Walls at Dholavira, Gujarat, India', *Current Science*, vol. 111, no. 12, 2016, pp. 2040–2043, p. 2040. https://doi.org/10.18520/cs/v111/i12/2040-2043.
8 S. R. Rao, *Dawn and Devolution of the Indus Civilization*, New Delhi: Aditya Prakashan, 1991, p. 392.
9 R. S. Bisht, *Excavation at Dholavira (1989–90 to 2004–05)*, 2015, pp. 921–922, asi.nic.in/asi_exca_2007_dholavira.asp.
10 M. C. Petersen, *Aggressive Architecture: Fortification of the Indus Valley in the Mature Harappan Phase*, Ph.D. thesis, Leiden University, 2012, p. 77.
11 M. S. Mate, 'Harappan fortifications: A Study', in *Studies in Indian Archaeology, Felicitation Volume*, ed. S. B. Deo, and M. K. Dhavalikar, Bombay: Popular Prakashan, 1985, pp. 75–83.
12 N. H. Hashimi, R. Nigam, R. R. Nair and G. Rajagopalan, 'Holocene Sea-level Fluctuations on Western Indian Continental Margin: An Update', *Journal of Geological Society of India*, vol. 46, 1995, pp. 157–162.
13 D. M. Maurya, M. G. Thakkar, A. K. Patidar, S. Bhandari, B. Goyal and L. S. Chamyal, 'Late Quaternary Geomorphic Evolution of the Coastal Zone of Kachchh, Western India', *Journal of Coastal Research*, vol. 24, 2008, pp. 746–758; S. B. Shukla, S. P. Prizomwala, V. Ukey, N. Bhatt, and L. S. Chamyal, 'Coastal Geomorphology and Tsunami Hazard Scenario along the Kachchh Coast, Western India', *Indian Journal of Geo-Marine Science*, vol. 39, 2010, pp. 549–556.
14 R. Nigam, R. Dubey, R. Saraswat, Sundaresh, A. S. Gaur and V. J. Loveson, 'Ancient Indians (Harappan Settlement) were Aware of Tsunami/Storm Protection Measures: A New Interpretation of Thick Walls at Dholavira, Gujarat, India', p. 2041.
15 James Rattue, *The Living Streams: Holy Wells in Historical Context*, Woodbridge: Boydell Press, 1995, p. 10.
16 E. J. H. Mackay, 'SD Area: The Great Bath and Adjacent Buildings', in *Mohenjo-daro and Indus Valley Civilization*, ed. John Marshall, London: Arthur Probasthain, vol. 1, 1931, p. 131.
17 Gregory L. Possehl, *The Indus Civilization- A Contemporary Perspective*, New Delhi: Vistar Publications, 2012, pp. 186–190.
18 John Marshall, 'The Buildings', in *Mohenjo-daro and Indus Valley Civilization*, ed. John Marshall, London: Arthur Probasthain, vol. 1, 1931, pp. 24–25.
19 Sina Panja, 'A Re-analysis of "Houses": The Evidence from Inamgaon, Maharashtra', *Bulletin of the Deccan College Post-graduate and Research Institute*, vol. 58/59, 1998–99, pp. 93–104; H. D. Sankalia, Z. D. Ansari, and M. K. Dhavilkar, 'Inamgaon: A Chalcolithic Settlement in Western India', *Asian Perspectives*, vol. 14, 1971, pp. 139–146.
20 Richard Sennett, *Flesh and Stone: The Body and City in Western Civilization*, New York: W.W. Norton, 1994, p. 139.
21 *Ṛigveda*, VII.47.1.
22 *Ṛigveda* I.32.95, V.19.7.
23 *Ṛigveda* X.121.1.
24 *Yajurveda*, III.6.

25 *Ṛigveda*, VII.95.2.
26 *Ṛigveda*, VI.49.7, 61.4.
27 *Ṛigveda*, X.64.9.
28 *Ṛigveda*, X.129.
29 *Ṛigveda*, X.75.
30 *Śatapatha Brahmana*, VI.8.2.2.2, III.6.1.7, IV.4.3.15.
31 Eliade, *Patterns in Comparative Religion*, p. 191.
32 Ananda K Coomaraswamy, *Yakṣas*, New Delhi: Munshiram Manoharlal Publishers, 2001, pp. 13, 24, reprint.
33 Gustav Oppert, *On the Original Inhabitants of Bharatavarsha or India*, Westminster: Archibald Constable & Co., 1894, pp. 24, 67–68.
34 *Chaṅdogya Upaniṣada*, VII.10.1.
35 Sudipta Sen, *Ganga the Many Parts of a River*, New Delhi: Penguin Random House, 2019, pp. 53–55.
36 Sen, *Ganga the Many Parts of a River*, pp. 55–57.
37 Sen, *Ganga the Many Parts of a River*, pp. 46–47.
38 Arpita Chattaraj (Mukhopadhyay), 'Sacred Water and Cultures of Worship: Some Observations on the River in India', *Humanities Bulletin*, vol. 4, no. 2, 2021, pp. 173–174.
39 Diana L. Eck, 'Ganga: The Goddess Ganges in Hindu Sacred Geography', in *An anthology of Writings on the Ganga: Goddess and River in History, Culture, and Society*, ed. Assa Doron, Richard Barz, and Barbara Nelson, New Delhi: Oxford University Press, 2015, pp. 233–251.
40 Jeffrey W. Jacobs, Lewis Owen and Gilbert F. White, 'Mekong River', *Encyclopedia Britannica*, 6 Mar. 2023. https://www.britannica.com/place/Mekong-River.
41 Edward Gargan, *The River's Tale: A Year on Mekong*, New York: Vintage Books, 2002, pp. 22–23.
42 Pathon Hongsuwan, 'Sacralization of the Mekong River through Folk Narratives', *Manushya: Journal of Humanities*, special issue, no. 19, 2011, p. 34.
43 *Suttanipāta Commentary*, II, pp. 407, 437–439.
44 *Final Report of Assessment of Current Status of Cauvery River in Karnataka and Study of 300 M Buffer Zone on Both Sides of the River*, published by Environmental Management and Policy Research Institute (EMPRI), Bengaluru, 2017, pp. 1–2.
45 Eliade, *Patterns in Comparative Religion*, p. 191–193.
46 B. Farrington, *Greek Science*, Harmondsworth: Penguin Books, 1963, p. 37.
47 W. K. C. Guthrie, *A History of Greek Philosophy The Earlier Presocratics and the Pythagorean*, Cambridge: Cambridge University Press, vol. 1, 1979, pp. 34–38.
48 Biswas, *History of Hydrology*, pp. 80–81.
49 Biswas, *History of Hydrology*, pp. 26–28.
50 Biswas, *History of Hydrology*, pp. 29–30.
51 Biswas, *History of Hydrology*, p. 32.
52 H. Gomperz, 'Problems and Methods of Early Greek Science', *Journal of the History of Ideas*, vol. 4, 1943, pp. 161–176.
53 Biswas, *History of Hydrology*, pp. 38–39.
54 Y. Karunadasa, 'The Buddhist Conception of Mahābhūtas as Primary Element of Matter', *University of Ceylon Review*, vol. XXII, no. 1–2, 1964, pp. 28–47.
55 B. Jowett, *Dialogues of Plato*, London: Oxford University Press, vol. 2, 1892, pp. 259–265.
56 Biswas, *History of Hydrology*. pp. 45–47.
57 Jamie Linton, *What is Water, The History of Modern Abstraction*, Vancouver/Toronto: UBC Press, 2010, p. 4.
58 W. K Nuttle, 'Is Ecohydrology One Idea or Many?', *Hydrological Science Journal*, vol. 45, no. 5, 2002, pp. 805–807.

59 Gregory L. Possehl, *The Indus Civilization, A Contemporary Perspective*, New Delhi: Vistar Publication, 2002, pp. 104–105.
60 B. M. Pande, 'Ring Wells in Ancient India', *Bulletin of the Deccan College Post-Graduate and Research Institute*, vol. 25, Silver Jubilee Volume, 1966, p. 207.
61 Pande, 'Ring Wells in Ancient India', p. 216.
62 *Vrhatsamhitā*, 54, pp. 121–122.
63 Biswas, *History of Hydrology*, p. 18; H. E. Wulf, 'The Qanāts of Iran', *Scientific American*, vol. 218, 1968, pp. 94–105.
64 A. W. Moore and John F. Terry, 'Water and Well-Worship in Man', *Folklore*, vol. 5, no. 3, 1894, pp. 228–229.
65 F. McCormick, 'Struell: Bathing at Midsummer and the Origins of Holy Wells', in *Religion, Cults & Rituals in the Medieval Rural Environment*, ed. C. Bis-Worch and C. Theune, Ruralia XI, Leiden: Side Stone Press, 2017, pp. 70–71.
66 Rattue, *The Living Stream: Holy Wells in Historical Context*, pp. 21–30.
67 B. Colgrave and R. A. B. Mynors, *Bede's Ecclesiastical History of the English People*, Oxford: University Press, 1969, pp. 107–109.
68 R. S. Bisht, *Excavation at Dholavira (1989–90 to 2004–05)*, 2015, p. 921, asi.nic.in/asi_exca_2007_dholavira.asp.
69 Possehl, *The Indus Civilization*, pp. 69–70.
70 D. R. Patil, *The Antiquarian Remains of Bihar*, Patna: K.P. Jayaswal Research Institute, 1963, p. 152.
71 Anand Singh, 'Giriyaka Hilltop Buddhist Monastic Complex (Rajgir): Understanding Antiquarian Remains and Physical Spaces', *Ancient Asia*, vol. 12, no. 13, pp. 1–19. https://doi.org/10.5334/aa.251.
72 *Epigraphia Indica*, vol. XX, pp. 71–72.
73 *Epigraphia Indica*, vol. XX (1929–30), pp. 71–72.
74 *Epigraphia Indica*, vol. VIII (1905–1906), pp. 45–49.
75 *Corpus Inscriptionum Indicarum, Girnar Inscription of Skandagupta*, vol. IV, pp. 82 ff. VV.26–39.
76 Julia Shaw, 'Archaeology of Buddhist Propagation in Ancient India: "ritual" and "practical" Models of Religious Change', *World Archaeology*, vol. 45, no. 1, pp. 83–108. https://doi.org/101080/00438243.2013.778132.
77 Anand Singh, 'Ecological and Environmental Concerns Amongst the Buddhist Communities of Ladakh: A Case Study', pp. 1–13. pp. 1–3. http://www.undv.org/vesak2011/panel3/11AnandSinghFINAL.pdf.
78 C. Matschoss, *Great Engineers*, London: G. Bells and Sons, p. 5.
79 Biswas, *History of Hydrology*, p. 2.
80 Georg A. Schweinfurth, *Auf unbetretenen Wegen in Aegypten*, Berlin: Hoffman und Campe, 1922, pp. 213-220.
81 G. W. Murray, 'Water from Desert: Some Ancient Egyptian Achievements', *The Geographical Journal*, vol. 121, 1955, pp. 173–175.
82 R. Payne, *The Canal Builders*, New York: The Macmillan, 1959, pp. 13–15.
83 G. A. Sarton, *History of Science: Ancient Science through the Golden Age of Greece*, Massachusetts: Harvard University Press, 1959, p. 79.
84 Biswas, *History of Hydrology*. pp. 14–15.
85 W. Phillips, *Qataban and Sheba: Exploring the Ancient Kingdoms on the Biblical Spice Routes of Arabia*, London: Victor Gollancz Ltd, pp. 200–201.
86 Biswas, *History of Hydrology*. pp. 68–69.
87 Karl A. Wittfogel, *Oriental Despotism: A Comparative Study of Total Power*, New Haven and London: Yale University Press, 1967, p. 3.
88 Wittfogel, *Oriental Despotism*, p. 15.

89 John W. Bennett, 'Anthropological Contributions to the Cultural Ecology and Management of Water Resources', in *Man and Water: The Social Sciences in Management of Water Resources*, ed. L. Douglas James, Lexington: University Press of Kentucky, 1974, pp. 35–36.
90 J. M. Donahue and B. R. Johnston, 'Conclusion', in *Water, Culture, and Power: Local Struggle in Global* Context, ed. J. M. Donahue and B. R. Johnston, Washington, DC: Island Press, 1998, pp. 339–346.
91 Vasudha Narayanan, 'Water, Wood, and Wisdom: Ecological Perspectives from Hindu Traditions', *Daedalus*, vol. 30, no. 4, *Religion and Ecology: Can the Climate Change*?, 2001, pp. 193–194.
92 Vasudha Narayanan, 'Water, Wood, and Wisdom: Ecological Perspectives from Hindu Traditions', p. 179.

2 Water in Buddhist Teachings and Monastic Life

People's prerequisite to sustain their life and anxiety for survival compelled them to consider water sacred and consequently, they gave due representation to water in their sacred pantheons as spirits and deities. People venerated and symbolised water in their prayers, ritual manifestations, and metaphysical assumptions. Water and its resources in different natural states like rain, vapour, snow, ponds, lakes, wells, rivers, and oceans have intrinsic values of regeneration, protection, and fertility. It covers 71 percent of the total area of planet Earth with 2.5 percent freshwater, 1.75 percent ice and snow, 0.74 percent groundwater, and 0.1 percent water in lakes and rivers. The water cycle moves through precipitation, evaporation, condensation, infiltration, and surface flow. The pattern of use of freshwater at the global level suggests that 70 percent of water goes to irrigation, 20 percent for industrial use, and 10 percent for domestic use for drinking, cooking, washing, and sanitation.[1] Though water is in abundance, it is not shared equally in different regions. Some areas have plenty of water; others lack water for drinking and basic sanitation. This tremendous natural divide always encourages people to search for and protect the water and its natural sources. Ancient civilisations conserved the water bodies and treated them as sacred. Vedic and Śramanic religions realised the importance of water in their life and developed different ideas and rules to venerate and consume water. Buddhism evolved its own metaphysical and cosmological views on water, where their teachings of impermanence and middle path became the central theme. The cosmological and corporeal values of water developed in the Harappan Civilisation and the Vedic period were accommodated in the larger context of Buddhism. Like other ancient religions, they realised the importance of water in human life and always endorsed a friendly approach towards it. Buddhism shares a special relationship with water and water resources. The springs, lakes, pools, rivers, and oceans are given due importance in Buddhist literature, and some are worshipped and considered the abodes of the Bodhisattvas. Water is also exemplified in their meditative means and therapeutical attributes. The Buddha emphasised reverence for water as it is vital for life. He often advises monks, nuns, and laypersons to look after water sources like rivers, ponds, lakes, and springs and take inspiration from them.

DOI: 10.4324/9781003601241-2

Water Symbolised in Buddhist Cosmological Traditions

Water's corporeal and cosmological significance on human life is inordinate and paramount. The Harappans were the first to understand water's cosmological significance in the Indian subcontinent. The sacredness of water is well exhibited in the archaeological remains of the Great Bath found in Mohenjodaro. The structure is considered an epitome of ancient times.[2] The metaphysical and cosmological orientation of *Ṛigvedic* Āryans is found in the tenth *maṅdala* of the *Ṛigveda*. Benimadhab Barua suggests that water cosmology developed in the *Ṛigveda* may be influenced by the Harappan Civilisation, which has developed a water cult and probably also metaphysical characteristics. He cites two *Sūktas*, the *Aghamarsana* and the *Prajāpati Parmeṣṭhiṅ* of the *Ṛigveda* where he found the cosmological attributes of water at par with the Saṅkhya and other Śrāmanic traditions. The *Aghamarsana* hymn explains how the world is created from *tapas* (heat) with the help of primordial water. The Prajapati Parameṣṭhiṅ endorses that from water all original substances, being and non-being are formed. Parameṣṭhiṅ can be considered a naturalist whose concept of nature was dynamic. The commonality of the context in the Harappan Civilisation, Vedic, and Śramanic periods suggests some interlinking and exchanges of ideas among the *Ṛigvedic* thinkers, the predecessors, and the successors.[3] The *Bhūmicala Sutta* informs that water is the nuclei of the Earth. At Capala Cetiya, Vaiśālī, when the Buddha decided to enter the *mahāparinibbāna*, a great earthquake occurred. When Ānanda enquired about the reasons for earthquakes, the Buddha told him there are eight reasons. The great earth rests on water (*mahāpathavi udake patistthita*), the water on the wind, and the wind on space. When the wind blows strongly, it interrupts the water, so the water shakes the earth. Again, when an ascetic or a Brahman has developed the insight, or when a god has cultivated the belief that the earth is insignificant in comparison to water which is immeasurable, then he disturbs the earth, shaking it, causing it to tremble, and quake. Again, when one is determined to be awakened, deviates from the company of the Divines, and consciously descends into his mother's womb, the earth is disturbed and quakes. Again, when one who resolves to awake mindfully and with full awareness leaves his mother's womb, the earth is disturbed; it shakes, shudders, and quakes. Again, when the Tathāgata awakens to the unsurpassed complete awakening, the earth is disturbed, it shakes, tremors, and quakes. Again, when the Tathāgata turns the wheel of Truth, the earth is disturbed and shakes, trembles, and quakes. Again, when the Tathāgata mindfully and with full awareness gives up the force of life and takes *mahāparinibbāna*, the earth is disturbed, shaking, shuddering, and quaking. Again, when the Tathāgata attains Nibbāna and becomes devoid of any remnant of attachment, then the earth is disturbed, it shakes, tremors, and quakes. These are the eight reasons for earthquakes.[4] Buddhism accepts that water is the key source of fertility and creation. Few beings are born on Earth, most are born and reside in the water.[5] Water sustains the majority of creators within its vast horizon and gives impetus to life and growth to all biotic and abiotic species on earth. It is an integral part of human life to grow and sustain biologically, aesthetically, and culturally.

Canonical and non-canonical literature informs that many monks and nuns eulogised the importance of rain, water, clouds, rivers, and oceans in their path to *arahanthood*. The *Pānīya Jātaka* narrates the story of six persons who became Pacceka Buddhas by repenting for their wrongdoings and by cultivating truthfulness. The *Jātaka* derives its name from one of them, a native of Kāsi, whose name was inspired by water. He travelled with his friends, each carrying a flask of water. On the way, when they felt thirsty, they drank water from each other's flask to save the water in their flasks. But soon, they started realising their misdeeds and Pānīya confessed his wrongdoings. His truthfulness earned him the status of a Pacceka Buddha.[6] Kāludāyī, on the instruction of Suddodhana, went to invite the Buddha to Kapilvastu. He joined the Saṅgha and later became an *Arahanta*. In the rainy season, when the rain started, covering the earth with an abundance of green plants and flowers, Kāludāyī requested the Buddha to visit his kinsmen, singing nature's splendours, mainly of rain in a series of *gāthās*. The Buddha accepted his request and visited Kapilvastu.[7] Bhikkhu Culasiva, a monk from Śri Lanka was not able to realise *jhāna* on impermanence and decided to go to Jambudvīpa from Mahātittha. On the way, he chose the ocean as his subject of meditation (*apo kasina)* and attained *arahanthood*.[8] The *Dona Sutta* says that as a lovely white lotus is not soiled by the water, the Buddha is unaffected by the material world. The Buddha says to Dona that as blue, red, or white lotus lilies, though born in the water and developed in the water, rise above the water and are left unsoiled by the water, even so, though born in the world and grow up in the world, I have overcome fruitless desires of the world and stay without any *saṁsāric* attachment. Remember me, Dona, I am the Buddha.[9] The *Milindapañha* mentions water as a purificatory and regenerative element that can be used as a stimulus to attain enlightenment. It helps to remove the internal factors of ignorance and defilement. When a drop of water falls on the lotus flower, it flows in different directions, disperses, scatters, and disappears. It does not stick to the lotus leaves. This phenomenon can be an example for monks and nuns developinginner purity to wither away all the riddles of life as water drops disappear from lotus leaves. It takes another example from the ocean. An *Arahant* should be like an ocean, which can assimilate all kinds of water from different sources but always emerge clean, removing all types of dirt and maladies.[10] Nāgasena apprises King Menander:

> *Just O'King as the lotus, though it is born in the water, and grows up in the water, yet remains undefiled by the water; just so, O King should the strenuous Bhikkhus, earnest in effort remain undefiled by the support that he receives, or by the followings of disciples ......that the first of qualities of lotus that he ought to have.*

And again O'King the lotus remains lifted up far above the water; just so O king, should the strenuous Bhikkhu, earnest in efforts remain above the worldly things.....

And again, O King, as the lotus trembles when blown up by the slightest breeze, a strenuous just so, O king should the strenuous Bhikkhus, earnest in the effort, exercise self-control in respect of at least of the evil dispositions, perceiving the danger (in the least offense).[11]

Nāgasena again gives examples of lotus, water, and ocean to define the attributes of *Nibbāna*. As the lotus is unaffected by the presence of water, the *Nibbāna* is not affected by any evil disposition. As the water is cool and eases the heat, the *Nibbāna* removes all kinds of evil dispositions. As water quenches the thirst of men and beasts and removes the craving for water, in the same manner, *Nibbāna* allays the craving for lust, future life, and material prosperity.[12] The Pali *Gāthā* texts also give several examples where water is exemplified as a simile to praise the Buddha and attain *arahanthood*. Udayin Thera praises the Buddha:

As the flower lotus
Arise in water, blossoms
Pure-scented and pleasing to the mind
Yet is not drenched by water
In the same way, born in the world
The Buddha abides in the world
And like the lotus by water
He does not get drenched by the world.[13]

The *Abhidhamma Piṭaka* conceives water an essential element of the universe. It is among the four *mahābhūtas-apo* (water), *vayo* (wind), *pathavi* (earth), and *tejo* (fire). These four created the universe and in the process of realisation of *jhāna*, become ultimate realities with their specific characteristics. The *apo* has the attribute of trickling and acting like a solidity that cannot be experienced by physical essence but only through the mind.[14] Buddhist literature categorises the form of water or its manifestations and signifies them as internal and external factors affecting human metabolism. Internal manifestations of bile, phlegm, pus, blood, sweat, fat, tears, serum, saliva, mucus, synovial fluid, urine, or other liquid. Sāriputta says when one sheds tears or leaves saliva, it shows that the elements are devoid of the self. The external manifestations of water are sometimes aggressive, which can lead to the destruction of villages, towns, and other regions or the huge quantities of water sieved beneath the earth and disappear. Both these forms are liable to change and are impermanent.[15] Buddhaghosha says that the mysteries of the *mahābhūtas* act like a magician who hypnotises the person to show how water is transformed into ice or an ordinary metal is converted into gold, but it does not happen. The person is more attached to ice or gold, but the reality is different. There is no ice or gold but the manifestations of the different *rūpas* that will fade with time.[16] The *mahābhūta* aims to identify the real *rūpa* to inculcate an understanding of impermanence and *anatta*. Water also symbolises an object to attain *jhāna*. It is one of the ten meditational totalities (*kasina*).[17] The *Viśuddhimagga* compares *pīti* (joy) of *jhāna* with water. It gives a gradual sense of attaining *jhāna* and *pīti* in different stages like minor (*khuddaka*), momentary (*khanika*), showering (*okkantika*), uplifting (*upbega*), and all-pervading (*pharana*) like a mountain. It is like a rock cavern swamped by a great mass of water.[18] The *Viśudhimagga* also informs that water *kasina* can be performed to realise *arahanthood*. It says:

> *Monk should fill a bowl or a four-footed water pot to the brim with water uncontaminated by soil, taken in the open through a clean cloth [strainer], or with any other clear unturbid water. He should put it in a screened place on the outskirts of the monastery as already described and seat himself comfortably. He should neither review its colour nor bring its characteristic to mind. Apprehending the colour as belonging to its physical support, he should set his mind on the [name] concept as the most outstanding mental datum, and using any among the [various] names for water (ápo) such as "rain" (ambu), "liquid" (udaka), "dew" (vári), "fluid" (salila), he should develop [the kasina] by using [preferably] the obvious "water, water."*[19]

The *Satipatthāna Sutta* of the Chinese *Madhyama Agama* says that a monk can develop rapture and pleasure out of seclusion, which pervades and fills his body. It is like the work of a skilled bath caretaker who fills the container with enough water, arranges bath soap, and makes water drench. It makes the whole surroundings pleasant and harmonious. In the same way, a monk develops happiness and bliss in his seclusion.[20] It is not only the Theravāda but also Mahāyāna and Vajrayāna that also praise water as a precious source of the universe and a great medium to attain Bodhisattvahood.

Mahāyāna speaks of only one great ocean (*Bhutatathāta*) and many waves (personalities). The real essence of each wave (the water) remains with other waves. The divine ocean is omnipresent and all-pervading; consequently, the universe subsists in it. Each drop of a divine ocean has an overwhelming effect on human life and reflects the exact nature of the ocean itself.[21] The Pure Land Buddhism enumerates the sea of Dharma of all Buddhas where they live and meditate. With their realisation of truth and strenuous efforts, they can eradicate ignorance, greed, and anger. The Buddha is like a lion, the most valiant of all men, and his glorious virtue is par excellence.[22] The ocean of the Tathāgata's wisdom is unfathomable, enormous, and boundless and its depth is known to him only.[23] The lotus and water connection has been a symbol of purity. The mud nourishes the root that symbolises the *saṁsāra*. When a flower rises above the water, the roots and stems remain in mud, showing the state of Buddhahood. The Buddha's nature and his wisdom are like the ocean. His *samādhi* is like a paramount mountain and his light of wisdom is brilliant and pure that outshines the sun and moon. Mahāyāna Sūtra not only eulogises the life and qualities of the Buddha and the Bodhisattvas but also emphasises the attributes and qualities of human life. Life and its interconnectedness are mirrored as the multicolours of blooming lotus that radiates its distinct lustre and shows the splendour of the enlightened realm. It is considered the realm of Pure Land, the world of enlightenment. Taitestu Unno examined the *Pure Land Sūtras* and found that lotuses with different colours blossom in a pool and provide a sublime environment for enlightenment. He says:

On the surface of the pools,
There are lotus blossoms as large as cartwheels
They are blue coloured with blue sheen
Yellow coloured with yellow sheen
Red coloured with red sheen
White coloured with a white sheen
They are delicate and fragrant.[24]

Pure Land Buddhism contemplates the different attributes of water from the perspective of living traditions in the monasteries. It can be used for corporeal purposes and also act as a symbol to meditate. Numerous legends are mentioned where the Buddha talks to Ānanda about the sublime quality of water and the qualities of the ponds. Some of the monasteries have a preponderance of ponds of enormous dimensions. Some are ten, 20, or 30 *yojanas* in length, width, and depth. Some are immeasurable extending over a 100,000 yojanas in length and width. Their water is clean, fragrant, tasty, and healthy with eight excellent qualities. It again informs how the Bodhisattvas and Srāvakas enjoy the bliss of water.

There are golden ponds with beds of silver sand; silver ponds with beds of golden sand; crystal ponds with beds of beryl sand; beryl ponds beds of crystal sand; coral ponds with beds of amber sand; amber ponds with beds of coral sand; agate ponds with beds of ruby sand; ruby ponds with beds of agate sand; white jade ponds with beds of purple gold sands; purple gold pond with beds of white jade sand. Others are composed of two to seven jewels. On the banks of these ponds are sandalwood trees, whose flowers and leaves hang down and diffuse perfume everywhere. Heavenly lotuses, blue, pink, yellow, and white bloom profusely in various tints and tones, completely covering the surface of the water.[25]

Pure land Buddhism enumerates that water soothes the body and refreshes the mental health. Its wholesomeness eliminates the defilements. Clean and pure water looks formless and transparent and its sand particles are shiny and glowing, and their glittering gleam can be visible despite the enormous depth of water. The particles of water merge and ripples in the meandering shape and their waves flow peacefully and vibrantly with all its aesthetic rhythm. The ripples of water wave quiver, and reverberate the countless delightful sounds of the Buddha. People feel the reverberating vibrations of water as a melodious voice of the Dhamma and the others represent it as the Saṅgha. For some, it symbolises quietude, emptiness, no-self, *mahākarunā*, forbearance, courage, meditative attributes, interdependence, and repudiation of defilements. These elements facilitate the path *Nibbāna*.[26] Buddhist cosmology covers a wide range of characteristics of water as a pivot element of the universe, right from the creation of the world to exemplify its attributes for the attainment of *arahanthood*. Water is one of the four intrinsic elements of the universe. It is considered sacred and precious because it sustains and endures all beings and other elements of Nature.

Corporeal Virtues for Water in Buddhist Monastic Traditions

Buddhist literature describes water used for domestic purposes as *pāniya* (drinking water) and *paribhojniya*, used in routine works like washing, rinsing clothes, cleaning the house, and the monasteries.[27] The monastic tradition of Buddhism was conscious that water is precious and certain norms and practices must be evolved for its judicious use and conservation. The Buddha was conscious of the fact that water is a necessity for the survival of human beings. The monks and nuns wandering the Gangetic plain always need water to quench their thirst, as nine months of the year are summer or humid. For *parikkhara*, the bhikkhus and bhikkhunīs were not allowed to demand special foods and privileges, but no such prohibition for water was introduced in Buddhist monasticism. On many occasions, the Buddha and the retinue of the monks visited the village and requested water to drink. In the early phase of Buddhism, specific *Vinaya* rules were developed for proper water use. To offer water is merit in Buddhism. The *Saṁyutta Nikāya* informs that those who develop groves, build bridges, and maintain water shade are meritorious persons (*Āramaropā vanaropā yejanā setukaraka. Papañca udapanan ca yedadanti upassayam tesam divā ca ratto ca sada punna pavaddhati*).[28] Udakadāyaka saw the Bodhisattva taking a meal and offered his water vessel to him. Because of this meritorious act, he was first born in heaven and then in the human world and became an *Arahant* under the guidance of the historical Buddha.[29] In his previous birth, he offered his pot of water to the Padmuttara Buddha, and as a reward, he could get drinking water at any spot he visited.[30] When a monk visits the monastery at a distant place, the monks must arrange and ensure the *pāniya* and *paribhojniya*. The resident monks will inform the guests about the source of water in the monastery and their proper use.[31] In the Buddhist monastic tradition, an elaborate water management system was developed in all parts of the Indian subcontinent.

The *Vinaya Piṭaka*[32] informs that (*yo pana bhikkhu jānaṃ sappāṇakaṃ udakaṃ tiṇaṃ vā mattikaṃ vā siñceyya vā siñcāpeyya vā, pācittiyaṃ*), the monks should not use or drink water containing insects.[33] The water can be an abode for small insects or worms. In this case, the bhikkhu must avoid throwing such water on the ground. If he violates this rule, it is a *pācittiya*. He also considers it inappropriate to throw water in a pool or a river that contains such insects that fall accidentally into the water but cannot survive. No fault is being committed if such water is thrown on the ground or the grass. If a bhikkhu drinks or uses water containing living beings, and he knows that drinking, showering, washing a bowl, extinguishing a fire, etc., can kill that insect, he commits a *pācittiya*. He is also advised to prevent insects from settling or breeding in the water within a container. The water in the container should be changed regularly. Before drinking water that could contain small insects, it must be filtered. The Buddha also instructed the monks to respect water and grains while eating and washing utensils. The monks should not throw water containing rice or other edible material in the area of habitation during the washing of the used vessels (*na sasitthakaṃ pattadhovanaṃ antaraghare chaḍḍassāmīti sikkhā karaṇīyā*).[34] The leftovers should be in an appropriate place where beggars and birds could use them. The monks were also instructed not to hold a vessel of water or a dish bowl with a dirty hand (*na sāmisena hatthena pānīyathālakaṃ paṭiggahessamīti sikkhā karaṇīyā*).[35] However, a Bhikkhu does not commit wrongdoing by holding a pot of water or a dish with a dirty hand to clean them or a pot containing water to rinse the hands. The leftovers of the food should be thrown in the pond so that its living beings can survive.[36] The Buddha was also conscious of the fact that the water of the ponds should not be polluted and the safety of the lives of living beings residing in the ponds should be ensured.[37] The monks and nuns were instructed not to defecate, urinate, or spit in the water.[38] The *Vaccakuti Vatthu* informs that Buddha recommended proper cleaning after the call of nature and using a separate sleeper (*ācamanapadukā*) for going to the toilets.[39] Lambert Schmithausen observes that strict toilet practices are part of the *patimokkha* to embrace the correct behaviour and maintain the monastic tradition. It was considered improper to defecate in open spaces or on green grass as it can kill minute sentient beings. The rule is more stringent for nuns. The deification of green grass by them is considered the *pācittiya* offense. It may be possible that due to urine and excrement, the grass may dry, which is against the Buddhist belief of *ahiṁsa*.[40] The restriction against throwing food remains on grass or into the water was to save the life of a minute creature living on grass or water. The Buddha advised the monks and nuns to respect even the tiny living beings on the one hand while understanding the value of leftover food and the requirements of other living beings. He also promulgated rules against travelling during the rainy season because of possible injury to insects and worms coming to the surface during rain.[41] The Kośāmbī dispute between the two groups of monks, followers of the *Dhamma* and the *Vinaya* is one of the foremost examples showing sensitivity to the optimum use and conservation of water. Later on, the *Vinaya* specialists became *Vinayadhara, and the Dhamma specialists became Dhammadhara*. The *Kauśāmbiya Sutta*[42] informs that in the ninth year of the Buddha's enlightenment, a minor issue regarding water use in the toilet erupted in the Saṅgha. One Dhamma monk went to the toilet and left some water in the bucket. As per *Vinaya* rules, water cannot be left in the bathroom and it comes under the *pāccitiya* offense, the *Vinaya* specialists objected and demanded disciplinary action against the monk who had done wrong. The monk apologised for his unintentional mistake but the *Vinayadharas* were adamant. Despite the Buddha's intervention, the issue was not resolved. He left the monastery and went to Parileyakavana. Later on, the *Vinayadharas* realised their fault and apologised.

The monks were instructed to maintain hygiene at both the personal and monastic levels. They were required to chew tooth wood and clean their teeth properly, and they were also required to carefully clean their tongues. I-Ching provides a fascinating example of the use of tooth wood by the monks in Nālandā and other monasteries.

> *Tooth-wood is known as Dantakastha in Sanskrit and etymologically derived from the words daṅta- tooth, and kashthaa, a piece of wood. Its ideal shape is about twelve fingers (aṅgula) long and should not be less than eight fingers. It is chewed softly at one end and after that, it works like a toothbrush. If one unavoidably comes near a senior, while chewing the tooth wood, one should cover the mouth with the left hand. After the proper brushing and rubbing of teeth, the wood should be properly bifurcated, and the tongue by bent. When used, the wood must be washed and thrown away. A stick is taken out of a large piece of wood, or from a small stem of a tree, or a branch of an elm, or a creeper, if in the forest; if in a field, of the paper mulberry, a peach, a sophora japonica (Huai), willow-tree, or anything at disposal. The freshly cut sticks must be offered to others, while the dry ones are retained for one's own use. The younger priest can chew as he likes, but for the elders, it should be hammered.*[43]

It was a routine practice to clean the teeth wisely and sanitise them properly to clean grains stuck in the crevices of teeth which will make gum healthy. Sputum or saliva should not be swallowed and must be spit out and the mouth should be properly washed either with pea flour or with ash mixed with water. It will erase face oil. Pouring and rinsing the mouth twice or thrice with clean water is considered hygienic and a regular practice in Indian monasteries as well among the lay people. It is mandatory for every monk and nun. Water for cleaning can be taken in a vessel from a water tank, and hands can be cleaned at some isolated place, or near a drain, or the steps leading down. The water vessel and the mugs made of conch shells for pouring water must also be cleaned with pea flour, ash, earthen materials, and cow dung. In India, eating and drinking were not permissible before mouthwash.[44] Monastics and familiar people were expected to practice high sanitation and hygiene standards. It gives impetus to maintain health, clean their surroundings, and coexist with biotic and abiotic elements like water, trees, plants, etc.

The Buddha made it mandatory for the monks and nuns to use a water strainer for safe and clean drinking water. Water strainer (*parissāvana*) is a must for every member of the monastic community right from the novices to elders. It is one of the eight requisites possessed by the monastic community. It was generally made of cotton cloths and has many varieties – *danaparissavana* (strainer with handle), *katacchuparissavana* (perforated ladle), *ottsaraka*, and *domakaraka* (water pot with strainer). The purpose of a strainer is to avoid drinking polluted water and save the life of minute insects.[45] It was made of a small piece of clean cotton cloth. Later on, monks were allowed to carry and use a strainer fixed onto a ladle, a double strainer, or a water filter to get clean water. The Buddha also instructed the monks to carry the water pots, and in case of urgent requirement, the monk must share his strainer with another monk. If he denies sharing his strainer, then he is guilty of *dukkhata*. Without a strainer and water pot, the Bhikkhu should use the corner of his upper robe to strain water before drinking. It was a norm in the monasteries and in the typical households that potable or drinking water was kept separately in the vessels and water for other domestic use was kept in big jars or small cisterns specially meant for it. The generic names for vessels are *kuṅdi* and *kalaśa* and both are used for storing water for drinking and other domestic purposes. Earthenware, porcelain jars, and vessels made of metals like copper or iron are known. The monks must maintain hygiene and cleanliness, and water should be touched or carried only by clean hands. The water vessels should be kept in a hygienic condition. Bathrooms and toilets should have their separate water vessels. Necked vessels with a small opening to pour water were preferred for drinking. The opening must be covered with a lid to avoid insects and dust. The lid may be bamboo, wood, linen, or leaves. It is a routine to clean the vessel before getting freshwater. For carrying water vessels during travel, jar bags made of cotton with appropriate strings are standard in monasteries. It helps to carry the water safely and protects the water and the vessel from dirt and insects. The monks carried the bag containing a water vessel in a travelling bag hanging from their shoulders. It is a rule that during the journey, the monks must carry water jars, bowls, and robes in a bag hanging on their shoulders and an umbrella in their hands. He is also allowed to carry a water jar for domestic as well as a bag to carry additional sleepers.[46] It was the duty of the monks to examine the water every morning in jars, wells, ponds, or a river. The water vessel should be examined first.

> *After pouring about a handful of it, by inclining the jar, into a pure bronze cup, a ladle made of bronze, a conch shell, or a plate of lacquer-work, pour it slowly on a brick. Or, employing a wooden instrument*

> *made for this purpose, observe the water for some moments, shutting the mouth with the hand. It is likewise well to examine it in a basin or a pot. Insects even as small as a hair-point must be protected. If any insects are found, return the water into the jar, and wash the vessel with other water twice until no insects are left in it. If there is a river or a pond in the neighbourhood, take the jar there and throw away the water containing insects; then put it in fresh filtered water. If there is a well, use its water, after filtering it, according to the usual manner. In examining well water, after some has been drawn, observe it in a water vessel, taking about a handful of it in a bronze cup, as stated above. If there is no insect, then the water can be used through the night, and if any be found it must be filtered according to the process mentioned above.*[47]

I-Ching informs that Indian monks use fine white cloth for straining water; fine silk is generally used in China. The strainer is made of soft silk measuring about four feet (*Hu-ch'ih*). It was folded vertically with its edges, then doubled by folding both ends and stitching them together. It functions like a net to filter water. The cords are attached to its two corners along with loops and a stick measuring approximately one foot and six inches is attached to stretch it wide. Its two ends are fastened to posts, and the vessel is put under to strain water. When water is poured, the insects get filtered from the water. A monk must examine the water running through the strainer. If some insects remain in the water, they are purified through the same process. The water should again be examined even passing of a night and a monk should be assigned for this purpose. It is also essential that during the straining process, the monks be cautious and aware to protect the lives of even the tiny insects. There are many ways of safeguarding life while drawing water. There are different kinds of strainers to draw or filter water from wells, ponds, and rivers. Sometimes, a double filter is used, and double vessels are placed. Sometimes the insects are tiny and quickly escape from strainers, for that many folded cotton or silk strainers should be used. Knobbed vessels with broad faces should be used to fetch water from wells and ponds. It should be plunged in water more than once, so the tiny insects will escape the pot and pure water will come to the vessel. Senior monks of the monasteries were advised not to touch or use the water kept in the monastery to strain, and the filter to be used for straining. The novices' duty was to strain the water and clean the strainers. Even ordained monks should always use a strainer and drink strained water only. Their water jars and vessels should be kept hygienic, and one should not wander without a strainer.[48] The Buddha's heightened consciousness towards cleanliness and hygiene led to the framing of extensive rules for the monastic community and the ordinary people. It begins with brushing the teeth, cleaning the tongue, washing the mouth, taking a regular bath, and storing water with full precautions and cleanliness.

The Buddha was against the ceremonial symbolisation of water. On many occasions, he condemned the ritual bathing. In one of the earliest instances, he exhorted the Jatilas, who used to go for ritual bathing during dark hours in extreme winter. He taught them the middle path is the most righteous way to eliminate mental and physical dirt.[49] Pali literature mentions many instances where monks and nuns admonished the people who believed in ritual bathing and sacraments. The *Therīgāthā* mentions that Punnika challenged a Brahmin Sotthiya who used to go to the river for the ritual bath to get rid of his sins. She ridiculed him for how water can clean a person from his sins. Are frogs, tortoises, snakes, crocodiles, and other aquatic animals free from their sins? Can butchers, hunters, killers, and other criminals take a bath in the river and get rid of their sins? Only the path of the Buddha removes the ignorance, not the water. Sotthiya was impressed and joined the Saṅgha.[50] Patācārā, a great nun of her age, became *Arahant* by realising the image of water. She had a troubled life as her husband and two children died at a very young age. The Buddha consoled her and made her realise the inevitability of death. She became bhikkhuni, and one day, while washing her feet, she noticed how the water trickled in small and large waves just as the mortals died, either in childhood, middle age, or old age. The Buddha sent her a ray of glory and appeared before her, speaking and confirming her thoughts. When he had finished speaking, Patācārā became *Arahant*.[51] Sundirika Bhāradvāja used to go to the river to take a purificatory bath. Once he told the Buddha to take a bath in the same river. The Buddha admonished him about how water in a river can wash off anyone's sin. He told him:

Bahuka and Ahikakha
Gaya and Sundirika too,
Paya and Sarasvati
And the stream Bahumati
A fool may there forever bathe
Yet will not purify dark deeds.[52]

Water became a symbol for removing physical dirt and a source of inspiration for meditation. Its regenerative power helps to maintain physical health and also acts as a catalyst for the monks and nuns to meditate and become an *Arahant*.

Bathing Rules and Practices in Indian Buddhist Monasteries

The *patimokkha* mentions the rule for fair use of water, conservation of drinking water, and water for daily use like bathing, cleaning, etc.[53] Personal sanitation and cleanliness are considered essential. The *Vinaya* is prescribed for hot and cold water for baths and other purposes as per the season's requirements. The *saddhavihārikas* (monk community of a monastery) will ensure the availability of such kind of facilities in the monastery.[54] The *Vinaya* also speaks that the *antevasika* must take care of his teacher, and hot or cold water must be supplied for his bath as per seasonal requirements.[55] It was also a tradition that water was offered to wash their feet when the monks or nuns arrived after wanderings. When the Buddha was residing in the Nigrodhārāma near Kapilvastu, he was invited to inaugurate the newly built Saṅthāgāra of the Śākyans. He and his retinue of the monks were welcomed by pouring water on their feet.[56] The monasteries should have separate places for water sources like wells, ponds, etc. The *Vinaya Pitaka* informs that *senāsana* (bed and sitting accommodation) should not be built near the sources of potable and washing water (*na paniya samanta senasanam pabbhotetabbam. Na paribhojaniya samanta senasanam pabbhotetabbam*).[57] The *senāsana* within a monastery is a residency for the monks. The water source near it can disturb their daily routines like meditation, teaching, and learning with their teachers and co-dwellers.

The *Saṁyutta Nikāya*[58] mentions that water of a lake is righteous and virtuous for bathing. It removes the dirt from the body. The *Vinayas* of different Buddhist traditions inform that bathing practices and their normative functions are for body care. Its essential purpose is cleaning the body and mouth, toiletries, shaving hair, cutting off nails, etc. These routine practices are not highlighted in Buddhist texts and traditions.[59] People in the ancient period were fond of taking baths in the rivers. On festival days, women of good families come for water sports, spending the whole day in the river. The kings and their retinue are also mentioned to take a splurge in water sports.[60] Sāriputta is mentioned as taking a bath in the waters of River Acirāvatī.[61] It was a common practice in Indian tradition to take regular baths to clean the body, but rare occurrences are also found where particular days were decided for the monks to visit the baths. The bhikkhus are informed not to use water for any entertainment. Playing in the water is a *pācittiya* (*udake hasadhamme pācittiyaṃ*). When King Prasenjit saw a group of 17 monks playing water sports in the River Acirāvatī, he conveyed the matter to the Buddha who prohibited the monks from going to the river for water sports. In a river, lake, pond, etc., where the water level reaches at least the height of the eyes, if a Bhikkhu dives, floats, and swims, to amuse himself, or becomes infatuated with these water bodies, he commits a *pācittiya*. If he rides on a boat for pleasure, or by throwing or causing to float anything on potable water, or within any container with water, he commits a *dukkhaṭa*. But a sick bhikkhu entering the water for health reasons does not commit any wrong thing, and also, a bhikkhu is allowed to swim to cross a river that hinders his way.[62] The *pācittiya* rule of the *Dharmaguptaka Vinaya* gives an interesting story that in a royal bath, the monks used to spend a long time. Because of that, the members of the royal household took a bath. When the official narrated the incident to the King, the monks were restricted to visiting the royal pool only once in 15 days. Ann Heirman says that the monks were restricted because bathing in the royal pool was an act of vanity against the *Vinaya*.[63] It may be possible, as well as possible, that the monks were allowed to go fortnightly to bathrooms and pools shared by the royals and commoners. In Indian tradition, the bath is considered a social duty and the Buddha also accepted this fact. However, in public space, the monks and nuns were allowed to visit bathing *ghāts* at the appropriate time. The *Vinaya Pitaka* mentions many bathing *ghāts* of River Acirāvatī where the monks and nuns used to go for baths. On some of the *ghāts*, the courtesans used to come for the bath. The Buddha prohibited the nuns from going to those baths. Despite the prohibitory orders, the Chabbaggiyā nuns did not stop visiting there and punitive action was taken against them.[64] The *Vinaya* texts also inform that the Buddha permitted the monks to build bathhouses for health and sanitation.

Problems like indigestion caused by overconsumption of food, unwillingness to work, and lethargy can be removed by taking a bath. This therapy was recommended to the monks by great physician Jīvaka.[65] The *Dharmaguptaka Vinaya* gives measurements and shapes of bathhouses. It should be square, round, or octagonal, not in front of a living room. It should be in the corner of the monastery. The bathhouse should have a door and also an arrangement of proper ventilation. Its floor should not be made from earthen materials but from stones and bricks. Arrangements should be made for hot and cold water and a suitable place should be allocated for hanging the robes.[66] Water source is essential for the bathhouse; a well should be dug or water should be arranged from a neighbouring place and stored in earthen vessels. A novice or *sramnera* should check these facilities when his teacher is going to take the bath. It must be cleaned and dried. In the case of an elderly teacher, his disciple should assist in cleaning and drying his body, sweeping the bathroom floor, and refilling the vessels for the next person.[67] The *Sutra on Bathing Monks in the Bathhouse* (*Wenshi xiyu Zhong-sing jing*) informs how Jīvaka gifted a bathhouse to the Buddha and enumerated several benefits of bathing before the Lord and Ānanda. Then the Buddha told Jīvaka about the benefits of bathing and the meritorious deeds accrued by a laity who donated a bathhouse. Washing consists of seven objects – fuel, pure water, bath beans, revival ointments, pure ashes, tooth wood, and undergarments. The bathing pacifies the four elements, earth, water, fire, and wind. It also eliminates seven kinds of illnesses, such as wind problems, dampness, cold, and dirt. After bathing, the body becomes light and eyes sharp.[68]

For bathing, especially in public places or pools, several rules are framed to cover the body adequately, that is, one must place oneself on the curve of the river or under the shadow of the tree. The body can be covered by the bathing robes or rain robes. The *Vinaya Piṭaka* mentions that the bathing robe of the bhikkhunīs should have proper dimensions and measurements. It should be six-hands (approximately one foot) long, and two-and-a-half-feet wide.[69] I-ching mentions elaborate arrangements for bathing in Nālandā Mahāvihāra and says that bathing in India is different from China. He talks about the moderate weather of India where seasons are not harsh and snowfall is not known. The winters are mild and, in the summers, the heat is not intense. People did not feel heat waves even in the peak of summer. In all the seasons, the people of India take baths regularly, and they begin their daily routine after baths. Many of them even did not eat before their bath. There is plenty of water in their surroundings and the pools have abundant water. Digging the pond for a public cause is considered sacred and dutiful. He found that there are 20 or 30 bathing places of different sizes at every yojana depending upon the availability of water, local requirements, and purpose. Boundaries did not protect the ponds, but it was common practice in the Magadha region that sāla trees were planted on all sides of the pond, and trees surrounded some of the ponds to the height of about 40 or 50 feet. All these pools are clean and rain-fed. There are many small and large bathing ponds within and outside Nālandā. Time was fixed for the resident monks to take a bath and every morning a bell (*ghaṅta*) was sounded to inform the monks to go for the bath during the bathing hour. Every monk used to come with their bathing sheets and sometimes a hundred, or a thousand monks together went to the different ponds from their monasteries to take baths.[70] Nālandā has strict regulations for bathing in ponds and rivers. Monks' bathing sheet or towel should be five feet long by a foot-and-a-half wide and woven from fine cotton. It should be wrapped around the waist over the undergarment during the bath. It should be knotted as:

> *Draw out and take off the ordinary undergarment, and let both ends of the bathing cloth come in front. Then take hold of the upper corner of the left end with the right hand, and by pulling it up towards the waist let it touch the body; join this with the right end of the cloth; and twisting both together, push them in between the waist and the cloth itself. This is the way of putting on a bathing sheet. When a man is about to come out of the bathing place, he should shake his body and emerge from the water very slowly lest he should take out some insects adhering to the cloth. The rules concerning the manner of coming up to the bank are laid down in the Vinaya text.*[71]

When bathing in the monastery without going to the pond, the bathing sheet is used similarly, and another man pours the water on. The place of bath should have an appropriate enclosure for bathing. He also informs that bathing sheets of three feet or less, as usual in China, is not considered proper, and bathing without any cloth is indecent and against the *Vinaya*. The monks and the lay people should use a bathing sheet made of cotton measuring the length four times its breadth, as it will be suitable enough to cover the whole body. Such bathing practice is as per the teaching of the Buddha and respectful among man and devas. Such customs should also be applied when one takes a bath at night.[72] The Buddha endorses building a bath-house or a brick pond in an open place. A special sanitorium for therapeutic purposes is also considered meritorious. It was also a practice to apply oil on the body and feet for health. These practices are good for maintaining clear eyesight and keeping off the cold.[73] An inscribed seal found in Kumrahār informs the donation of the sanitorium to the Buddhist monastery for maintaining the good health of the monks.[74] It is common practice among the Indians and Buddhist monasteries that bathing should be done before a meal. The pouring of water and rinsing of the body removes the dirt and makes the body fresh. It also has medicinal purposes, as bathing helps to digest food and make the body healthy. In the ancient Indian medicine system (*Āyurveda*), bathing is prohibited after a meal, but bathing after the meal is allowed in China. The Chinese proverb says, 'Wash hair when hungry, but bathe after food, ' but this is not true in India.[75] Regular baths and washing the body are natural habits and customs in the Indian climatic conditions as they maintain good health and save people from infections and ailments. All religious traditions belonging to the Indic traditions, wholistically recommend regular bathing as obligatory, as it accentuates self-confidence, physical fitness, and heightened emotional well-being. The Buddhist monastics accepted the rich bathing culture prevalent in India from the period of the Harappan Civilisation as an obligatory practice for the monks and nuns. It removes dirt and odours from the body and encourages them to engage in meditative practices with heightened consciousness. Hand-washing was also a practice they embraced as the monks came into contact with many potentially harmful bacteria during their wanderings.

Vassavāsa (Abode in Rainy Season)

The Buddha was concerned with the safety of water sources as the lives of millions of sentient beings depend upon them. While framing the rules of *vassavāsa*, the Buddha shows concern about the life of sentient beings during the rainy season. The abode in the rainy season was an essential and universal mechanism in Indian ascetic traditions. Many of the popular sects in the Gangetic Valley during the age of the Buddha practiced *vassavāsa*. It was a practical as well as spiritual demand of the time. Due to heavy rain in the monsoon season, the middle and lower Gaṅgā region plains were heavily flooded and it wasn't easy to travel to distant places and for alms. Treading on muddy land could also harm the newly born tiny living beings and plants because the Indic ascetic traditions restrained themselves from wandering in the rainy season. It is believed that the Jains were the first who framed the rules not to travel in rainy seasons (*pajjusaṇa*) as their wanderings can trample newly grown grasses, plants, and insects. The *Acarāṅga Sūtra* informs that in the rainy season, Jain monks and nuns should not wander or tread on the path as many sentient beings are born and many seeds are sprouting. The wandering from one village to another could injure the green herbs, hurt the vegetable life, and destroy the lives of many nascent living beings.

> *When the rainy season has come and it is raining, many living beings are originated and many seeds just spring up, the roads between (different places) contain many livings, seeds, the footpaths are not used, the roads are not recognizable. Knowing this (state of things) one should not wander from village to village, but remain in during the rainy season in one place.*[76]

The beginning of *vassavāsa* in Buddhism was a chance incident. Once, during a heavy rain, the monks wandered the road, and the people showed annoyance and reacted.

> *How can the Sakkyaputtiya Samanas go on travels alike during winter, summer, and rainy seasons? They crush the green herbs, they hurt the vegetable life, they destroy the life of many small living beings.*[77]

When the monks informed the Buddha that people were annoyed, they murmured against the monks for wandering during the rainy season, the Buddha permitted the monks and nuns to spend the rainy season in one place. The foundation of the *Vassavāsa* (abode in the rainy season) in Buddhism indicates that people were conscious that insects and plants could be destroyed if not tread cautiously during the rainy season. The Buddha told the monks:

> *Monk! I Allow you to enter upon the rains in the rainy season* (*anujanāmi bhikkhave vassāne vassaṇupagantam ti.*)[78]

The *Mahāvagga* informs two periods of rain retreat: earlier (*pūrmika*) and later (*pacchimika*). The first begins with the Indian lunar month of *Aṣādha* full moon day and the latter just one month later in *Sāvana*. It lasted up to the full moon day of the month of *Assyuja* or *Kartika*. Later on, the Buddha also allowed the monks to take leave of seven days for specific reasons related to Saṅgha and Dhamma.[79] The Buddha spent 45 *vassavāsas*, the first 20 in the different parts of the middle Gaṅgā Valley and the last 25 in Srāvastī at Jetavana and Pubbārama. He spent his first *vassavāsa* at Deer Park in Sārnāth. The second, third, and fourth *vassavāsa* were in Rājagṛiha. The Buddha spent the fifth *vassavāsa* in Vaiśālī and resided in the Mahāvana. The sixth was spent in Mankulagiri. In the seventh, he went to the Tvatimsa heaven to preach Māyā Devī. The eighth *vassavāsa* was spent in the Bhesakala forest near Sumsumārgiri. The ninth and tenth *vassavāsa* were spent in Kosāmbī and Parileyyaka-vana, respectively. The 11th, 12th, 13th, 14th, and 15th *vassavāsa* were spent in Ekanāla, Veranja, Cāliya rock, Jetavana, and Kapilvastu. The 16th, 17th, and 18th were at Alavaka, Rajagrha, Cāliya Rock, and again the 19–20th at Rājagriha.[80] *Vassavāsa* led to the development and consolidation of Saṅgha in its nascent phase. A growing number of monks and nuns transformed the structure of the Saṅgha. The problems and impediments coming in that development were an ample opportunity for the Buddha to frame or change the rules, and develop teachings and practices that could endure Saṅgha for a longer duration. With the vagaries of time, many changes occurred in the regulations of *vassavāsa*. Different sects framed their own rules for the *vāsa*.

Vassavāsa was the beginning of a settled monastic culture and ushered in new cultural and academic vistas in Indian religions, notably Buddhism. Initially, Buddhist monks were not allowed to stay at a dwelling for more than three consecutive nights. There were no established monasteries and the monks wandered, stayed in a forest, at the root of a tree, on a hillside, in a glen, in a mountain cave, in a cemetery, in the periphery of the forest, in the open air, and on a stack of straw. The *Cullavagga* informs an incident where a merchant of Rājagriha once saw that the bhikkhus were without the settled abodes and were forced to stay in an open space, he expressed a desire to build dwellings for them. The bhikkhus did not consent because the Buddha had not permitted them to accept such dwellings. When the incident was reported to the Buddha, he allowed residency in five types of abodes, that is *vihāras*, *addhayogas*, storied dwellings, attics, and caves. After that, the merchants of Rājagriha built 60 dwelling places in one day and donated to the Saṅgha of the four quarters.[81] The residency in the months of the rainy season provided better acquaintances for the community of monks to inculcate the *Vinaya* and Dhamma in a more structured way. The lay Buddhist disciples could use that period to intensify their learning and practice of Buddhist teachings as they were more accessible to the monks and nuns. They can make merit by providing the requisites of the monks who are observing the rain retreat who in turn will teach them the Buddhist doctrines. So, the events during the rain retreat can be of shared benefit to both the monastic and the lay disciples. Buddhaghosa says that the monks who intend to spend three months at a particular monastery are expected to make a formal declaration – '*Imasmim vihare imam te-masam vassam upemi*' (entering a three-month rain retreat in this particular dwelling).[82] The purpose of the ordained monks to observe the rain retreat in one place was to circumvent any risks to themselves and to evade damage to crops and living beings coming on their paths during their wanderings. The rain residency provided an opportunity for them to learn and practice the scriptures and the disciplinary rules (*Vinaya*) and to deepen the practice of meditation. In the beginning, two kinds of dwellings – *āvāsa* and *ārāma* were established. *Āvāsa* was a temporary settlement and could be dismantled after completion of *vassavāsa* and *ārāma* was a more settled structure. During the *vassavāsa*, the bhikkhu had the right to an accommodation (*senāsana*) in an *āvāsa*. The Buddha made special rules for the dwelling and making of *āvāsa* and *ārāma*. It should not be too near or too far from a village or urban settlements.[83] It witnessed the institutionalisation of the important activities and ceremonies of the Saṅgha such as the *sīmā* (boundary), *uposatha* (fortnight assembly), *upasampadā* (final ordination), *pāvarana* (confession), *kathina* (offering of robes), and discussion on the teachings of Tathāgatna, etc.

Usually, the *vassavāsa* was followed by two ceremonies, viz, *pāvarana* and *kathina*. The *pāvarana* was a valued observance in which each bhikkhu had to confess his depravities of commission and omission, committed if anything happened during the *vassavāsa*. It is a common affair after the end of the abode in the rainy season and the monks are expected to inform and exchange the means and ways of what has been seen or heard. It was held on the full moon of the month Assyuja or Kartika. The Buddha told the monks that after the end of the rain retreat, the *pāvarana* should be convened to exchange words on what is heard or suspected. Such verbal exchanges and confessions will strengthen the bond among the members of the Saṅgha and the *patimokkha* will be followed and implemented more congenially.[84] It was a mechanism to understand and endure each other's behaviour as well as to forgive each other's transgression. *Pāvarana* was an expression of collective adherence to the *Vinaya* and a resumption of *carika* for the welfare of the whole world. The culmination of the *pāvarana* was the *kathina*. It denotes a robe offering to the monks after the end of the rain retreat. Etymologically, the word *kathina* means firm or hard and probably represents the wooden frame used for weaving these robes. The Buddha prescribes that those who have completed the first cycle of *vassavāsa* are eligible for robe offerings in the month between Assayuja to Kārtika. A monk can receive a robe only once. The monks who receive *kathina-civara* enjoy certain privileges, viz., free to go to the meal invitation without informing the fellow monk (*anāmantacāra*), enjoy the group meal with four or more monks (*ganabhojana*), etc.[85] The word *kathina* is challenging to define in early monastic terminology. However, it looks more appropriate to accept that robe offering symbolises a problematic path ahead that a bhikkhu has to follow till the next *vassavāsa* to enhance the spirit and cohesion of the Saṅgha.

Inspirations from Lakes and Rivers

Like other Indian religions, Buddhism also considers rivers a divine water source because of their curative characteristics, life-giving, and means of prosperity as water sustains our life, irrigates agricultural fields, and helps navigate large ships going for long-distance trade. Jambudvīpa is popularly known as an island because multiple rivers meander and traverse the very part of this region. The region is embedded with and surrounded by perennial water sources from the Great Himalayan hills. The visual and literary pieces of evidence show that people in the ancient past were conscious about water sources and the origin of rivers from the great hills. Buddhist literature states that the five greatest rivers originated from a lake formed by the Himalayas. The cosmological view is that the human world and other visible and invisible things have originated from the primordial nature of water. On its deeper levels, Buddhism throws a bright light on the subjectivity of hills, water sources, and their religious and worldly relevance. It reveals, with great pragmatism, how the Himalayas and their everlasting source of water are essential for divinity and societal cause and explains, as a consequence, that there are as many assumptions to understand the value of water resources. It hypothesises that there were heavy rains in the Himalayas due to the atmospheric action and all the water coming from the rain led to the formation of the seven lakes. These seven lakes are Anotatta, Kannamunda, Rathakāra, Chaddanta, Kunāla, Mandākinī, and Sīhappapāta. These lakes are surrounded by the five hills of the great Himalayas, that is Sudassanakūta,

Citrakūta, Kālakūta, Gandhamādana, and Kelāsa. G. P. Malalasekera summarises this process based on literary descriptions mentioned in the *Aṭṭhakathās*. He describes the formation of lakes as

> *Four channels open out of the lake in the direction of the four quarters: Sīhamukha, Hatthimukha, Assamukha, and Usabhamukha. Lions abound on the banks of the Sīhamukha; elephants, horses, and cattle respectively on the others. Four rivers flow from these channels; the eastward river encircles the lake three times, waters the non-human regions of Himavā, and enters the ocean. The rivers that flow north and westward flow in those directions through regions inhabited by non-humans and also enter the ocean. The southward river, like the eastward, flows three times around the lake and then straight south over a rocky channel for sixty leagues and then down a precipice, forming a cascade six miles in width. For sixty leagues the water dashes through the air onto a rock named Tiyaggala, whereon by the force of the impact of the waters the Tiyaggalapokkharani has been formed, fifty leagues deep. From this lake the waters run through a rocky chasm for sixty leagues, then underground for sixty leagues to an oblique mountain, Vijjha, where the stream divides into five, like the fingers of the hand. The part of this river which encircles the original lake Anotatta is called Āvattagangā; the sixty leagues of stream which run over the rocky channel, Kanhagangā; the sixty leagues of waterfall in the air, Ākāsagangā; the sixty leagues flowing out of the Tiyaggala-pokkharanī and through the rocky gorge is called Bahalagangā, and the river underground, Ummaggagangā. The five streams into which the river is divided after leaving the oblique mountain Vijjha are called Gaṅgā, Yamunā, Aciravatī, Sarabhū, and Mahī.*[86]

The Buddha has high reverence for Himalayan hills and their water sources. One reason may be that Lumbinī and Kapilvastu where Siddhārtha was born and brought up are situated in the Himalayan Terai in Eastern Uttar Pradesh and Nepal, swamped with lakes and ponds with enormous water sources. The water of the Anotatta Lake is considered pure, sacred, regenerative, and imperishable. Its water will dry only at the end of the world.[87] On the day of her conception, Mayā Devī saw the dream of a white elephant and was taken to the Anotatta Lake by the Devas to give her a divine bath.[88] It is also believed that the Buddha used to go to the Anotatta Lake for the bath and to Uttarakuru for alms, then return to the lake to have his meal and spend some time on its banks.[89] The Anotatta and the other four lakes were surrounded by the habitations of the great sages and wandering ascetics who engaged themselves to develop several metaphysical and cosmological notions. The area irrigated by it is considered not only the birthplace of the historical Buddha but also of a number of the early Buddhas, Boddhisattvas, and Pacceka Buddhas who came, resided, and meditated here. There are many bathing places in the lake where the Buddhas, Pacceka Buddhas, monks, ascetics, devas, and other inhabitants of the deva-worlds, and the goddesses used to come for baths.[90] The water from Anotatta was a sublime source of meditation. Sometimes, the Buddha would go there with monks to preach.[91] The perennial water system of the Himalayas and its lakes make this region highly fertile and dense forest with rich flora and fauna, creating an ideal environment for affluent material life and worldliness as well as a situation to wander an unworldly life.

Though the modern parameters of meteorology and geomorphology were not known in the age of the Buddha and other sages, they were wise enough to perceive how mountains, rain, and lakes play key roles in the formation of the river systems and the vast agrarian plains. They always followed and endorsed a kind of sacred bond between these water sources and human beings. The *Cullavagga* describes the five great rivers (Mahānadī) in the age of the Buddha: Gaṅgā, Yamunā, Acirāvatī, Sarabhū, and the Māhī.[92] The *Milindapañha* mentions 500 rivers that flow down from the Himalayas. Among them Gaṅgā, Yamunā, Acirāvatī, Māhī, Sindhu, Sarasvatī, Vetravati, Vitamsā, and Chandrabhaga are important ones.[93] Early Buddhist texts eulogise Gaṅgā, Acirāvatī, Anomā, Rohinī, Niranjanā, and Hirañyavatī because these were directly associated with the life of the Buddha. Gaṅgā has been mentioned as one of the five great rivers (Mahānadī) that waters the Jambudvīpa, the others being Yamunā, Acirāvatī, Sarabhū, and Mahī.[94] She is also known as Alakanandā, Bhāgirathī, and Jāhanvī. Alakanandā flows towards the Gaṅdhamadhana hills from the south of the Sumeru and it finally falls into Mānasarovara. The Bhāgirathī comes from Gangotrī in the Garhwal district to Haridvar and finally irrigates the middle Gaṅgā plain. From Haridwar to Prayāga, she flows parallel to River Yamuna and makes Sangam with Yamuna at Prayāga. From here to Rājmahal hill in Bengal, she meanders through an easterly course and finally merges as River Padmā with the Indian Ocean.[95] Because of the unending flow of water, some of the great ancient cities like Vārānasī, Champā, Prayāga, Pataliputra, and Saṅkissa are situated on the banks of River Gaṅgā. These cities played an important role in the formation and development of the Buddhist Saṅgha.[96] It is believed that Gaṅgā is meandering over the 500 leagues.[97] The Buddha has given numerous examples and similes on the name of Gaṅgā. The attributes of the Buddha are considered infinite as the sands of Gaṅgā.[98] The water of Gaṅgā looks full of foam but the foam is empty.[99] The water of Gaṅgā flows with mighty force and it is impossible to restrict its movement with one's fist.[100] The Gaṅgā and other rivers merge with the ocean like the monks should find solace only in *nibbāna*[101] The Buddha also teaches impermanence by giving an example of the abundant source of water of Gaṅgā. It is inevitable like the flow of water of Gaṅgā as it always meanders to merge with the sea.[102] The meeting point of the two great rivers of the north, Gangā and the Yamunā is often symbolised as an example of meditation and perfect union.[103] Buddhism is one of the first religions of India which emphatically endorses the curative and regenerative power of the water of Gaṅgā. The preaching of the *Ratana Sutta* along with the sprinkling of water of Gaṅgā to cure the people of Vaiśālī from epidemic became a *paritta*. This incident was celebrated even in Rājagriha as the Gangārohana.[104] The water of Gaṅgā was considered holy and was used for the consecration of kings in India as well as in Śri Lanka.[105] Sand particles are often examined for numerous characteristics of enlightenment. All Buddhas, tathagatas, and the Bodhisattvas in the ten directions as numerous as the sands of the Gaṅgā, together pray the supreme virtues of Amitayus.[106] The relevance of Gaṅgā in the mystical and spiritual realm among all Indic religions is everlasting. It also waters the most fertile region of India, making it self-sufficient and prosperous in agrarian terms.

Yamunā is the second of the five great rivers of Jambudvīpa.[107] As a great tributary of Gaṅgā, she rises from the Himalayas below Mount Kamet and meanders through the hills of Siwalik and Garhwal and then enters into the plains of Uttar Pradesh. During her movement, she meets with different tributaries like Tons, Karnavatī or Chambal, Kalīsindh, Vetravati or Betwa, Ken, and Payasunī.[108] In Buddhist literature, Yamunā is mainly referred to exemplify the conduct and attributes of the monks, nāgas, and lay persons. Bakkula's story is very popular. He was born into a wealthy family of Kośāmbī but drowned in River Yamunā while being bathed by his nurse. He was swallowed by a fish which was caught and sold by a fisherman to the family of Vārānasī. When the fish was cut, the child was discovered unhurt and brought up by them. Later on, he came to know about his parents and the king decided that the two families jointly should bring up and care for him. His name 'Bākkula' means split open or divided.[109] Bakkula joined the Buddhist Saṅgha and became *Arahant*. He was known for his psychic power (*iddhi*).[110] The water of Yamunā is also infested with nāgas.[111] Acirāvatī or modern Rāpti is also one of the five great rivers flowing from the Himalayas eastwards and finally merges into the sea.[112] The Jain texts call it Eravatī.[113] It carries extensive water from the Himalayas during the rainy season when it becomes flooded but in the summer seasons, this river turns dry leaving a vast bed of sand in the whole Himalayan Terai region.[114] During the age of the Buddha, it streamed through Kośala, and many royal edifices of King Prasenjta were built on its banks at Srāvastī.[115] The Buddha performed a great miracle to counter the heretics near the bank of Acirāvatī in Srāvastī.[116] He also preached the *Tevijja Sutta* where Acirāvatī is used in a simile to prove the futility of sacrifices and rituals.[117] When Vidudbha invaded Śākyas and killed thousands of people, it is believed that due to the Buddha's psychic powers, he was drowned in Acirāvatī along with his large army.[118] Two of the greatest vihāras, the Jetavana Vihāra and Pubbārama Vihāra built by Anāthapindika and Visākha, respectively, were situated at Srāvastī near River Acirāvatī. Mahī is mentioned in Buddhist literature as one of the five great rivers of India.[119] It has been identified with modern Gandak. Sarabhū is also one of the five great rivers of northern India.[120] It has been identified with modern Saryū. Saryū divided the north and south Kośala. Acirāvatī is a tributary of Sarabhū that passes through the Anjanavana, a famous Buddhist monastic site.[121] Hoey identifies Sarabhū as modern Saryū and the Māhī either a tributary of Gaṅdak or Gaṅdak herself. Acirāvatī has been identified as River Rapti.[122] Sarabhū merges with River Ghaghara, a tributary of the Gaṅgā. The famous city of Ayodhyā is situated on the bank of this river. Ayodhyā was the second capital of Kosāla *mahājanapada* and Srāvastī was the first. This river is considered sacred by the Ikṣvākus.

Anomā is considered one of the sacred rivers by Buddhists as here, Siddhārtha declared his intention to be a wanderer. It was flowing east of Kapilvastu, the capital of the Śākyans. The river has been identified as modern Aumi, an old bed of River Rapti.[123] Later Buddhist literature gives a different array of distances between Kapilvastu and Anomā. Siddhārtha rode the whole night on his favourite horse Kaṅthaka to reach Anomā from Kapilvastu.[124] On the bank of it, he cut off his hair and adorned the yellow robe of the ascetics. Charioteer Channa and his favourite horse Kaṅthaka were witnesses to that gracious moment. It is believed that Kaṅthaka intentionally muffled the sound of his neighing and footsteps as he galloped through the streets. Channa caught the tail of Kaṅthaka to follow him. Reaching outside the rampart of Kapilvastu, Siddhārtha stopped once to have a last look at the city. Kanthakanivatta-vihāra was built at that spot. The horse travelled 30 leagues between midnight to reach River Anomā. On arriving on the opposite bank, Siddhārtha gave orders that he should be taken back to Kapilvastu, but Kaṅthaka kept looking back at his master, and when he disappeared, Kaṅthaka died of a broken heart on the spot and was reborn in Tāvatimsa as Kaṅthaka-devaputta.[125] River Anomā became a symbol where Siddhārtha first time experienced the agony of transience and separation. His *mahābhiniṣkramana* was the end of his princely life and the beginning of life as a wanderer and an ascetic. He lost his favourite horse as well as separated from his whole family and Anomā was the sole witness of these miraculous moments.

Niranjanā or Neranjarā is considered sacred for many reasons. On the bank of it, Siddhārtha wandered and many great happenings occurred that led to his enlightenment. This river has been identified as modern Phalgu with its two branches Nelanjana and Mohana. It originates at Simeria, in district Hazaribag of Jharkhand. Bodhagayā, the place of Buddha's enlightenment is situated on the bank of this river.[126] The Pali *Aṭṭhakathās* says that the river is known as Niranjanā because its water is pure, pleasant, and blue.[127] The area was surrounded by lush green forests, and because of that, it was a source of attraction for the wanderers to come and meditate. The great Jatila ascetic Nadi Kassapa had a hermitage on the bank of this river.[128] Here Siddhārtha took austerities along with the *pañcavaggiya* bhikkhus. During his rigorous penances, he partook only one jujube fruit and a grain of rice. He meditates without any interruption but soon realises the futility of austerities, the *pañcavaggiyas* get angry and abandon him. He came to Uruvelā, on the banks of Neranjarā and here Sujatā, daughter of the chief of Uruvela offered him a bowl of milk rice. Siddhārtha bathed at the bathing *ghāt* Suppatittha and partook in the milk rice. After that, he spent some time in Sāla grove on the bank of this river before the night of the Enlightenment.[129] Here, on the bank of the river, probably he discovered the path of the middle way. After the Enlightenment, the Buddha resided under the Ajpala Nigrodha at Uruvela. Here, Māra tempted him to dissuade his noble path. Here, Brahma Sahampati persuaded him to preach the Dhamma.[130] The river witnesses the most prolific actions and happenings that shaped Siddhārtha into the Buddha. Even the experiences encountered here with the *pañavaggiyas* inspired him to make them his first disciples at Sārnāth.

Rohīnī is a small river flowing in the Himalaya Terai. It was a dividing line between two ancient republics, the Sākyas and Koliyas. People of both republics shared the water of this river for agriculture and irrigation. Once, in the encroaching summer in May (Jetthamūla), drought engulfed the whole region, and the water level of River Rohinī receded. People of both republics started quarrelling over the exact share of water for irrigation, but soon it turned violent and the battle seemed imminent. At that time, the Buddha was residing in Jetavana, when informed about the incident, he came to intervene and convince both parties about the foolishness of killing each other for a bit of water. He preached the *Phaṅdana*, the *Duddaba*, and *Latukika Jātakas*. Later, he also delivered the *Rukkhadhamma*, the *Vattaka Jātakas*, and the *Attadaṅda Sutta* to exhort the people of both republics. Both groups, the Sākyas, and the Koliyas, conciliated and offered 250 young men from each clan to join the Saṅgha.[131] The *Suttas* preached here became the *summons bonum* to resolve the water disputes and water crisis. The nations with a sizable presence of Buddhist population still rely on these teachings to resolve personal and political conflicts. Hiraṅyavatī has been identified as Choti Gandak which flows through the Gorakhpur region in modern Uttar Pradesh. It is flowing parallel to Badi Gandak. This river merges with River Ghaghra.[132] On the bank of this river, the Upavattana Sālavana of the Mallas was situated. The Buddha, on his journey of the *mahāparinibbāna*, took a bath in this river and crossed it to reach Kuśināra.[133] The journey of the Buddhahood that began from the bank of River Anomā, ended on the bank of Hiranyavatī. With his *mahāparinibbāna* at Kuśinara, the physical presence of the Buddha ended but from a religious perspective, his *śarīrakāyā* was still present in the form of his body relics. The banks of rivers like Gaṅgā, Acirāvatī, Niranjanā, Rohinī, and Hiranyavatī played significant contributions in the formation of two most critical Buddhist ideologies, that is *Śarīrakāyā* and *Dhammakāyā*.

Notes

1 A. Balasubramanian, 'The World's Water', 2015. https://www.researchgate.net/publication/315123891.

2 E. J. H Mackay, 'SD Area: The Great Bath and Adjacent Buildings', in *Mohenjodaro and Indus Civilization*, ed. John Marshal, London: Aruther Probsthain, 1931, p. 131.

3 Benimadhab Barua, *A History of Pre-Buddhistic Indian Philosophy*, Delhi: Motilal Banarasidas, 2016, pp. 15–16, reprint.

4 R. M. L. Gethin, *Sayings of the Buddha: A Selection of Suttas from Pali Nikāyas*, Oxford: Oxford University Press, 2008, pp. 61–62.
5 *Saṁyutta Nikāya*, V.467.
6 *Jātaka*, no. 459.
7 *Aṅguttara Nikāya*, I.25; *Theragāthā*, VV.527–536.
8 *Viśuddhimagga*, V.170.
9 *Aṅguttara Nikāya*, IV.36.
10 *Milindapañha*, IV.6.32–33.
11 *Milindapañha*, VII, 2.4.
12 *Milindapañha*, IV.66.
13 *Theragāthā*, VV.700–701.
14 *Viśuddhimagga*, XI.93.
15 *Aṭṭhaśālinī*, vol. II, III.335.
16 *Viśuddhimagga*, XI.98.
17 A. K. Warder, *The Path of Discrimination* (*Patisambhidamagga*), London: Pali Text Society, 1982, pp. 379–380.
18 *Viśuddhimagga*, IV.94–99.
19 *Viśuddhimagga*, V.3.
20 E. Ciurtin, 'The Buddha's Earthquakes, On Water: Earthquakes and Seaquakes in Buddhist Cosmology and Meditation, with an Appendix on Buddhist Art', *Stvdia Asiatica*, vol. X, 2009, p. 80.
21 Wm Montgomery McGovern, 'Notes on Mahayana Buddhism', *The Monist*, vol. 29, no. 3, 1991, pp. 381–382.
22 Hsao Inagaki, *The Three Pure Land Sutras, The Larger Sutra on Amitayus, The Sutra on Contemplation of Amitayus; The Small Sutra on Amitayus*, Berkely: Numata Centre for Buddhist Translation and Research, 1995, p. 9.
23 Inagaki, *The Three Pure Land Sutras*, p. 35.
24 Taitestu Unno, *River of Fire, River of Water: An Introduction to Pure Land Tradition of Buddhism*, London: Double Day, 1998, p. 8.
25 Inagaki, *The Three Pure Land Sutras*, p. 26.
26 Inagaki, *The Three Pure Land Sutras*, p. 27.
27 *Vinaya Piṭaka*, II, 207; IV, 263.
28 *Saṁyutta Nikāya*, I.70.
29 *Apadāna*, I, p. 354.
30 *Theragāthā Commentary*, I.115.
31 *Cullavagga*, VIII.1.
32 *Vinaya Piṭaka*, IV.48–49.
33 *Vinaya Piṭaka*, IV.125.
34 *Vinaya Piṭaka*, IV.199.
35 *Vinaya Piṭaka*, IV.198.
36 *Aṅguttara Nikāya*, I.16.
37 *Majjhima Nikāya*, I.13.
38 *Vinaya Piṭaka*, IV.350.
39 *Vinaya Piṭaka*,, II, p. 222.
40 Lambert Schmithausen, *Problem of Sentiency of Plants in Earliest Buddhism*, Tokyo: The International Institute of Buddhist Studies, 1991, pp. 31–36.
41 Anand Singh, *Plants Planet and Animals, Ecological Paradigms in Buddhism*, New Delhi: Primus Publishers, 2019, p. 50.
42 *Majjhima Nikāya*, 48.
43 J. Takakusu, *A Record of the Buddhist Religion as Practiced in India and the Malay Archipelago, AD 671–695*, New Delhi: Munshiram Manoharlal, 1998, pp. 33–34.
44 Takakusu, *A Record of the Buddhist Religion as Practiced in India and the Malay Archipelago, AD 671–695*, pp. 26–27.

45 *Cullavagga*, V.13.2.
46 Takakusu, *A Record of the Buddhist Religion as Practiced in India and the Malay Archipelago, AD 671–695*, pp. 27–30.
47 Takakusu, *A Record of the Buddhist Religion as Practiced in India and the Malay Archipelago, AD 671–695*, p. 31.
48 Takakusu, *A Record of the Buddhist Religion as Practiced in India and the Malay Archipelago, AD 671–695*, pp. 31–33.
49 *Vinaya Piṭaka*, I.81, *Udāna*, 9.
50 *Therrgāthā*, VV.236 51.
51 *Therrgāthā*, VV.112–116.
52 *Majjhima Nikāya*, 121.
53 *Patimokkha*, no. 3.
54 *Vinaya Piṭaka*, V.212–213.
55 *Vinaya Piṭaka*, III.116.
56 *Saṁyutta Nikāya*, IV.200.
57 *Vinaya Piṭaka*, II.213.
58 *Saṁyutta Nikāya*, I.182.
59 Ann Heirman and Mathieu Torck, *A Pure Mind in a Clean Body, Body Care in Buddhist Monasteries of Ancient Indian and China*, Gent: Academia Press, 2012, p. 5.
60 *Papancasūdanī*, II.604; *Dhammapadaṭṭhakathā*, III.199.
61 *Manorathapūranī*, I.315.
62 *Vinaya Piṭaka*, III.1–2.
63 Heirman and Torck, *A Pure Mind in a Clean Body*, p. 28.
64 *Vinaya Piṭaka*, IV.259ff.
65 *Vinaya Piṭaka*, II.119–122.
66 Heirman and Torck, *A Pure Mind in a Clean Body*, pp. 31–32.
67 *Vinaya Piṭaka*, I.47; II.224.
68 Heirman and Torck, *A Pure Mind in a Clean Body*, pp. 33–34.
69 *Vinaya Piṭaka*, IV.278–279.
70 Takakusu, *A Record of the Buddhist Religion as Practiced in India and the Malay Archipelago, AD 671–695*, p. 108.
71 Takakusu, *A Record of the Buddhist Religion as Practiced in India and the Malay Archipelago, AD 671–695*, p. 109.
72 Takakusu, *A Record of the Buddhist Religion as Practiced in India and the Malay Archipelago, AD 671–695*, pp. 107–110.
73 Takakusu, *A Record of the Buddhist Religion as Practiced in India and the Malay Archipelago, AD 671–695*, pp. 127–129.
74 K. K. Thaplyal, *Studies in Ancient Indian Seals*, Lucknow: Akhil Bhartiya Sanskrit Parishad, 1972, p. 215.
75 Takakusu, *A Record of the Buddhist Religion as Practiced in India and the Malay Archipelago, AD 671–695*.
76 *Acarānga Sūtra*, p. 136.
77 *Mahāvagga*, III.1.
78 *Mahāvagga*, III.3.
79 *Mahāvagga*, III.4.
80 *Narada, A Manual of Buddhism*, Kuala Lumpur: Buddhist Missionary Society, 1992, pp. 64–66.
81 *Mahāvagga*, III, 4–5.
82 T. W. Rhys Davids and Herman Oldenberg, *Vinaya Texts*, SBE, Vol. XIII (I), Delhi: Low Price Publication (reprint), p. 299.
83 *Vinaya Piṭaka*, I 39.
84 *Mahāvagga*, IV.4–5.
85 *Mahāvagga*, VII.1.

86 G. P Malalasekera, *Dictionary of Pali Proper Names*, New Delhi: Motilal Banarasidas, 2007, vol. I, pp. 96–97.
87 *Aṅguttara Nikāya*, IV.101.
88 *Papañcasūdanī*, II.918.
89 *Vinaya Piṭaka*, I.28.
90 Malalasekerea, *Dictionary of Pali Proper Names*, vol. I, pp. 97–98.
91 *Apadāna*, I.299.
92 *Cullavagga*, IX.1.3.
93 *Milindapañha*, IV.1.35.
94 *Vinaya Piṭaka*, II.237; *Saṁyutta Nikāya*, II.135.
95 Bimla Churn Law, *Historical Geography of Ancient India*, New Delhi: Munshiram Manoharlal, 2016, pp. 77–78.
96 Malalasekera, *Dictionary of Pali Proper Names*, vol. I, p. 734.
97 *Saratthapakāsinī*, II.119.
98 *Saṁyutta Nikāya*, IV.376.
99 *Saṁyutta Nikāya*, III.140.
100 *Saṁyutta Nikāya*, IV.298.
101 *Majjhima Nikāya*, I.493.
102 *Saṁyutta Nikāya*, IV.179.
103 *Jātaka*, no.412, 415.
104 *Dhammapadaṭṭhakathā*, III.444.
105 *Mahāvaṁsa*, XI.30.
106 Inagaki, *The Three Pure Land Sutras* Pure, p.31.
107 *Vinaya Piṭaka*, II.237; *Aṅguttara Nikāya*, IV.101, 198.
108 *Law, Historical Geography of Ancient India*, pp. 135–136.
109 *Theragāthā Commentary*, I.344.
110 *Viśuddhimagga*, 379.
111 *Dīgha Nikāya*, II.259.
112 *Vinaya Piṭaka*, II.237.
113 *Kalpasūtra*, p.12.
114 *Aṅguttara Nikāya*, IV.101.
115 *Vinaya Piṭaka*, IV.111–112; *Suttanipāta Commentary*, I.19.
116 *Dhammapadatthakatha*, III.205; *Mahavastu*, III.115.
117 *Dīgha Nikāya*, I.235–253.
118 *Dhammapadatthakatha*, I.360.
119 *Vinaya Piṭaka*, II.237; *Aṅguttara Nikāya*, IV.101.
120 *Vinaya Piṭaka*, II.237; *Udāna*, V.5; *Saṁyutta Nikāya*, II.135; *Aṅguttara Nikāya*, IV.101.
121 *Aṅguttara Nikāya*, I.104.
122 W. Hoey, 'The Five Rivers of the Buddhists', *The Journal of Royal Asiatic Society of Great Britain and Ireland*, 1907, pp. 44–46.
123 Alexander Cunningham, *The Ancient Geography of India*, New Delhi: Munshiram Manoharlal, 2016 (reprint), pp. 484–485.
124 *Vimānavatthu Commentary*, V.314.
125 *Buddhacarita*, V.3, 68; VI.53ff.
126 *Law, Historical Geography of Ancient India*, p. 244.
127 *Udāna Commentary*, pp. 26f.
128 *Theragāthā Commentary*, I.45.
129 *Dhhammapadaṭṭhakathā*, I.71.
130 *Vinaya Piṭaka*, I.1ff; *Saṁyutta Nikāya*, I.103f; 122, 136; *Aṅguttara Nikāya*, II.20f.
131 *Suttanipāta Commentary*, I.358; *Dhammapadaṭṭhakathā*, III.254ff.
132 *Law, Historical Geography of Ancient India*, p. 85.
133 *Dīgha Nikāya*, II, 137.

3 Water in Buddhist Religious Practices

Parittas and Holy Water

Water is sacred and intrinsic to religious beliefs. It is healing, purifying, and regenerative. Water as a symbol of purity and a source of regenerative attributes has been worshipped and used across societies since the foundation of human civilisations. Different religions used this idea according to their traditions and practices. The Buddha also understood the importance of the regenerative powers of water and linked it to human health and meditation. The custom of holy water in Buddhism goes back to the age of the Buddha and the Pāli sources suggest that it may have the power to give well-being, drive away diseases, put evil spirits to flight, etc. It is believed that the Buddha himself developed this sacrament that a monk blesses by reciting a particular *Sutta* with a sprinkling of holy water from his alms bowl, imploring the Buddha's blessings on those who suffer from any problem. It also ensures protection from the powers of evil forces. Water also becomes sacred when it is poured into the image of the Buddha on the occasion of celebrating his birth. The practice of this cherished liturgical ceremony with familiar Pali chants is still popular across the Buddhist world. It is considered sacred fully embedded with spiritual power and can combat evil spirits through the Buddha's divine grace.

Water Ceremony on the Birth of Siddhārtha

The sprinkling of water to celebrate the birth of the Bodhisattva Siddhārtha is a popular cult in the Buddhist tradition. The bathing ceremony begins with sprinkling holy water thrice on the image of the infant Bodhisattva. It visualises the aim to remove impurities accrued by greed, anger, and ignorance. Donald K. Swearer says that Buddha abhiṣeka is a popular ceremony in Theravāda tradition and its references are found in the Buddhist canonical and non-canonical texts. In the consecration ceremony, an elaborate ritual is performed to eulogise the birth of Siddhārtha, his quest for enlightenment, and his benevolent acts. These enumerations are found in the *Buddhavaṃsa*, *Cariyapitaka*, *Jātaka*, *Nidānkathā*, *Buddhacarita*, and *Lalitavistara*. The ceremony is considered to be a post-canonical phenomenon, but the portrayal of the multicoloured connotations of the Theravāda tradition. It is a modern

DOI: 10.4324/9781003601241-3

predilection; the ceremony makes the Buddha a radiant source of energy and bestower of the world.[1] It is believed that the holy water ceremony was first performed when Queen Mahāmāyā gave birth to Siddhārtha. Before this event, ritual bathing was known among the Śākyas. When Queen Mahāmāyā conceived, she dreamt that the Devas had taken her to the Anotatta Lake to give her a bath. At birth, Siddhārtha was given ritual bathing as per the royal practices of the Śākyas.[2] The *Buddhacarita* informs how the divinities accorded the Bodhisattva Siddhārtha royal bathing.

> *As soon as he was born the thousand-eyed (Indra) well-pleased took him gently, bright like a golden pillar; and two pure streams of water fell down from heaven upon his head with piles of Mandāra flowers.*[3]

This practice is celebrated annually in the Indian lunar month Vaiśākh on full moon day (Vesak day) to symbolise the inner purification and purgation of sins. The pouring of holy water on the child Buddha embodies calmness and serenity. It reminisces the episode that the Devas were showering rain on the newly born Siddhārtha. Water is considered the universal solvent, which can wash away greed and avarice. The legend enumerates that Bodhisattva was born mystically. His infant body was clean and free from blood and mucus. His mother is shown standing in the Lumbinī forest holding the branch of the Sāla tree when she delivered Prince Siddhārtha. It is believed that just after the birth, he strode seven steps, and at this movement, below his feet, a lotus sprang up.

The *Acchariyābhūta Sutta*[4] informs that when infant Siddhārtha came out from Mahāmāyā's womb, he was clean, unsoiled by water, mucus, blood, or any other kind of impurity, and two streams of water appeared in the sky, one cool, one warm, for bathing of the infant. After birth, he stood firm on the ground. Facing north, he takes seven strides with a white parasol above him, sees all quarters, and exhorts: 'I am the foremost in the world! I am the eldest in the world! I am the first in the world! This is my last rebirth. Now there are no more future lives.' The perfumed water poured on the image of the infant Buddha, showing his right forefinger pointed upwards and left forefinger pointed towards the earth, becomes blessed. It is a celebration with holy water to bring sentient beings with a promise to end suffering and ignorance. Water sprinkling connects the cosmos of the Buddha, monks, and laity. The monks and laity remember the Buddha and the other Buddhas, Bodhisattvas, his family members, and associates. They are the bestowers of good looks, prosperity, and health. Even the opposing forces of Māra are also remembered to be watchful against *akuśala kamma* and other kinds of defilements (Figure 3.1).

Water reflects change and transition. It also visualises relativity and illusion. The performance of the water ceremony reflects the protection and prosperity of the people and also indicates the illusory attributes of human perceptions about the material world. With its generative prowess, water

Figure 3.1 Bodhisattva Siddhārtha after his birth (Lumbinī), Image by author.

reflects the vision of transcendence and enlightenment. The association of lotus and lotus pool with the Buddha endorses the notion that as the lotus comes out from mud and blossoms, the Buddha in the same way is free from all *saṁsāric* bondage, and his path is the best to remove suffering and ignorance. The blossoming of the lotus flower shows the Buddha's nature and the inexhaustible water source symbolises *Nibbāna*. Water is also the source of hope, wisdom, and enlightenment. The Buddha as a ferryman is a real torchbearer and guide who helps humans cross the *bhavasāgara* embedded with impermanence and suffering and attain the *arahanthood*.[5] The *Buddhacarita* reiterates the symbolism of how holy water became linked to the auspiciousness of his birth.

> *When he was born, the earth, though fastened down by (Himālaya) the monarch of mountains, shook like a ship tossed by the wind; and from a cloudless sky there fell a shower full of lotuses and water-lilies, and perfumed with sandalwood.*
>
> *Pleasant breezes blew soft to the touch, dropping down heavenly garments; the very sun, though still the same, shone with augmented light, and fire gleamed, unstirred, with a gentle lustre.*
>
> *In the northeastern part of the dwelling, a well of pure water appeared of its own accord, wherein the inhabitants of the gynaeceum, filled with wonder, performed their rites as in a sacred bathing place.*[6]

Mahāyāna developed elaborate rituals for the Buddha's bathing ceremony. The '*Sūtra on the Merit of Bathing the Buddha*' a text dealing with the ceremony was first written in the precincts of Nālandā Mahāvihāra and copied by Chinese traveller I-Ching in the 7th century. It was translated into Chinese for the first time by an Indian monk Manicintana in 705 CE and then by I-Ching in 710 CE. Its Sanskrit and Tibetan versions are not known. A similar kind of text '*Sūtra on the Merit of Building Stūpa*' was translated into Chinese by the Indian monk Divakara in 680 CE.[7] The *Sutra* was developed in the early medieval period to pay homage to the Buddha in two forms, the worshipping of the living Buddha and the veneration of the Buddha after his *mahāparinibbāna*. The ceremony for the first kind of worship included offering incense, flowers, food, and water to the Buddha, and pouring water on his image to accrue unlimited merit and blessings. The merit earned in this ceremony even can lead to enlightenment. The text specifies that the worshipper also pays homage to the body relics of the Buddha along with a water ceremony. Boucher refers to the Sūtra-

> *Although the great teacher is extinguished, his image is still present, one should venerate it with an elevated mind as if the Buddha is still here. Some may place incense and flowers (before the image) every day enabling them to produce a pure heart. Others may perform the bathing ritual, completely cleaning their tenebrous karma. Those who apply their thoughts to this practice will automatically receive their unlimited reward. As for those who exhort others to do it, the merit they have already produced will be compounded. Those who seek blessings should set their minds on this. Moreover, all the monasteries of the Western region (especially India and Central Asia) bathe the noble image.*[8]

I-Ching vividly describes the ceremony of bathing in the image of the Buddha in the Indian monasteries. In the forenoon, the monk in charge of the ceremony (*Karmadāna*) rings a bell (*ghaṅta*) to inform the inmates of the same. The image of gold, silver, copper, or stone is put in the basin and water from perfumed water jars is poured on it. When water is poured on the head of the image, it brings good luck and the ceremony has immense reward.[9] The Sutra on Bathing the Buddha says that the *abhiṣeka* ceremony gives a multitude of merits, prosperity, protection, and fortunate rebirth. When Bodhisattva Pure Wisdom asked the Buddha why and how the bathing of the Buddha's image would accrue merit? The Buddha replied that it would create a platform for right mindfulness. He says:

> *After my death*
> *You will be able to honour my relics*
> *Some will build stupas*
> *Or images of Tathagata*
> *At the place of the image or stupa*
> *One who anoints that spot of ground*
> *With various incense and flowers*

Scattering them over its surface,
Uses pure, beautifully scented water
To pour onto the body of this image
Offers it various flavourful drinks and foods,
Fully maintaining it with oblations,
Eulogizes the virtue of Tathagata.
Which is endlessly difficult to conceive
Through the wisdom of skillful means and the supernatural power (of the Buddha)
Such a one will quickly reach the other shore (of nirvana)
He will obtain the diamond body
Complete the thirty-two marks of a great person
And the eighty minor signs of excellence
He will ferry the multitude of living beings (to the shore of nirvana).[10]

I-Chings mentions that on the fast day water ceremony is also performed and it was well celebrated by the laities and Indian monasteries were well engaged in it. The laity prepares a mixture of betel nut, aromatic oil extracted from *Fu-tzu* (*mustaka* or *Cyperus rotundus*), and a small amount of crushed rice and puts it on a leaf on a plate. The mixture is covered with a white cloth and put into a golden vessel. A vessel with the mixture is sanctified with water. After that, the monks are invited to the ceremony in the forenoon. The monks anoint their bodies, wash, and bathe. In the midday of the second day, the image of the Buddha was carried outside the monastery on a carriage or a palanquin to the layman's house, and the procession was accompanied by a multitude of priests and laymen, playing drums and musical instruments. They make offerings of incense, flowers, and unfurl flags that shine in the sun. When the procession reaches its destination, the image of the Buddha made of gold or bronze, covered by an umbrella, is anointed with some aromatic paste, and then put in a clean basin. It is bathed by all those present with perfumed water (*gaṅdhodaka*). After being wiped with a scented towel, the image is carried into the principal hall of the house, where it is received amidst rich offerings of lights and incense. The Sthavira recites the *Dānagāthā* for the host to accrue the merit for him and his family. Then the monks are offered food and gifts. After receiving food, they wash their hands and rinse their mouths. Finally, they are offered sugar water (squash) and Pin-lang fruits (i.e. betel nuts).[11] The ritual bathing given to the Bodhisattva became symbolised as holy water. It is developed as a safety mechanism for the well-being of the Buddhist fraternity. Interestingly, the monastic community and laity are covered under the broader spectra of sacredness and sanctity of holy water. The Buddha's annual birth celebration is a reminder of the purpose of the arrival of the Bodhisattva. Its scope was extended when sprinkling of water became a sacrament along with chanting of the *Suttas* to avoid danger and inculcate wellbeing. It also reminds the followers of the Buddha to fulfil their duty with honesty and the Tathāgata will take care of all beings. It has different contexts in the different Buddhist traditions. In the Mahāyāna countries, a child image of the Buddha (baby Buddha) is generally used for bathing ceremonies during Vesak day. The Theravāda tradition links these images and bathing ceremonies with different pagodas/stūpas. In the most sacred Shwedagon Pagoda of Myanmar, each day, different images of the Buddha, Bodhisattvas, and their associated mounts were assigned for the water sprinkling ceremony. A layperson can visit the image of the respective day in the surroundings of the stūpa and pour water on it. Special utensils and water tabs were arranged for this purpose in front of images of the Buddha. In Thailand, the Buddha's life and its linkages with its different social groups are personified by different images. The image of the Emerald Buddha in the Royal Temple of Bangkok is associated with royalty and represents the King. The Lion Buddha image of Chiang Mai is the bestower of rain and prosperity. In the annual New Year festival (*Songkran*), the people use to take this image in procession for proper rain and blessings.[12] Donald K. Swearer examined the water sprinkling (Buddha – *abhiṣeka*) ceremony in Thailand and found:

> *The consecration ceremony establishes a common thread of meaning inherent in most Buddhist rituals by connecting the founder with past and present, the dead and living. Two designations given to the ritual illustrate this generalized significance: "eye opening" (boek phranet) and "consecration" (abhiseka). To open the eyes of a Buddha image is to enliven it, to bring it to life, to make it present, to instill it with power. Within the context of Buddhist ritual and ceremonial practice, to abhiseka means to consecrate by pouring water or lustrating. At the deepest level, abhiseka not only makes sacred in the sense of purifying but it also recreates and make new as symbolized by the life-giving form of water.*[13]

The water pouring on the image of the Buddha connects the laity, that is lokiya, to the Buddha who is supermundane (*lokuttara*) with the intention that remembrance of the Buddha will always bring happiness and bliss.

Parittas and Holy Water

The mechanism of safety through verbal action, water sprinkling, and sacred thread tying was developed in Buddhism from its beginning and is called *paritta*. Paritta is derived from the Sanskrit root '*pari*' and '*tra*' means to protect. It was a mechanism of chanting the *Sutta* with a sprinkling of water to ward off all kinds of trouble. It is proposed that the *paritta* was developed to deal with the psychological requirement of the familiar people who desired to derive some mystical powers to cope with the day-to-day sufferings of their life. The lay followers could not understand the deep knowledge required for *Nibbānic* pursuit. The chanting of the *Sutta* gives them a firm belief in the Buddha. The water and thread used in recitations provide a psychological satisfaction that miraculous powers are operating to protect them from all kinds of harm.[14] The beginning of this tradition in Buddhism is linked to the occurrence of the snake biting a monk and the ghastly epidemic taking the lives of humans in Vaiśālī. Once, the people of the Vajji Saṅgha were troubled by famine, epidemic, and drought, the Lichchavi King narrated the unfortunate incident of the unnatural deaths of their innocent citizens to King Bimbisāra of Magadha and requested him to placate the Buddha to visit Vaiśālī to save the lives of innocent people. When the Buddha was informed of the incident, he decided to go to Vaiśālī. This episode gave birth to a mechanism of the '*paritta*' or safety device using water. The Buddha preached the *Ratana Sutta* on this occasion. The Pāli literature reflects many *parittas* for protection against epidemics, snake bites, and physical harm. Lily de Silva[15] divides the *parittas* into two broad categories: prophylactic and benedictive. The prophylactic *parittas* are (i) to facilitate protection against dangers, recovery from sickness, and epidemics; (ii) to dissipate fear; (iii) for safe childbirth; (iv) safety from wild animals; (v) protection in battle; (vi) to avoid spirits, ghosts, or the impact of hostile planets; (vii) to free the person affected by a spirit; (viii) protection from natural disasters like floods, drought, feminine, etc.; (ix) to avoid worldly troubles as predicted by astronomical calculations.

The important *parittas* in the benedictive category are (i) to get positive vibrations; (ii) *paritta* for earning merit; (iii) before beginning the *vassavāsa*; (iv) before starting a journey; (v) before the beginning of social functions like marriage, birthday, etc.; (vi) beginning of the legislative sessions; (vii) to ensure righteous governance; (ix) world peace and just world order; (x) ensuring good rain; and (xi) praying for the longevity of life etc.

Beginning of *Paritta* of 'Blessed Water'

The *Ratana Sutta* is explained in the *Khuddakapātha* and the *Suttanipāta* in 17 verses. The first two verses are dedicated to the beings residing in terrestrial or celestial regions. The Buddha develops a bridge between humans and divine or semi-divine spirits to live in harmony. The humans will offer them homage and in reciprocation, the divine beings will protect them from danger. The third verse exhibits the supremacy of the Buddha and his inordinate qualities. In the following verses, the virtues of the Buddha, the Dhamma, and the Saṅgha are discussed. In the last, the celestial beings under the leadership of Sakka express their respect and admiration for the Buddha, Dhamma, and Saṅgha. The first three stanzas[16] of the *Sutta* are:

1 *Yānīdha bhūtāni samāgatāni Bhummāni vā yāni va antalikkhe Sabbeva bhūtā sumanā bhavantu Atho pi sakkacca sunantu bhāsitam.* (Whatever beings are assembled here, terrestrial or celestial, may they all have peace of mind and may they listen attentively to (my) words.)

2 *Tasmā hi bhūtā nisāmetha sabbe Mettam karotha mānusiyā pajāya Divā ca ratto ca haranti ye balimTasmā hi ne rakkhatha appamattā.* (Listen here, all beings. May you all shower loving kindness to those human beings who bring offerings to you by day and night. Therefore, protect them with diligence.)
3 *Yam kiñci vittam idha vā huram vā Saggesu vā yam ratanam panītam Na no samam atthi tathāgatena Idam pi Buddhe ratanam panītam Etena saccena suvatthi hotu.* (Whatever treasure there be either here or in the world beyond, whatever precious jewel there be in the heavenly worlds, yet none is comparable to the Enlightened One. This precious jewel is the Buddha. On account of this truth, may there be well-being.)

The observation on the mechanism of the *Ratana Sutta* gives many threads showing its purpose, methodology of recitation, disseminating agency, parties engaged on the occasion of the recital, endurance for the Dhamma and Saṅgha, and finally the omnipresence of the Buddha. The occasion of developing the water ceremony preached in the *Ratana Sutta* was linked to a horrific problem in Vaiśālī, the capital of the Lichchavis. The city was ravaged by continuous famine, epidemic, and unnatural deaths. The Lichchavis were so disturbed that they sent his commander Mahāli accompanied by one of their priests to request King Bimbisāra of Magadha for the Buddha's intervention to resolve the problem. The *Mahāvastu* informs that 1000 sons of Yakṣini Kuṅdala spread the deadly disease. They spread two kinds of epidemics, *maṅdalaka* and *adhivāsa*. The first covers the whole family and the second consists of the whole area of Vaiśālī. The Lichchavis sent Tomara to Rājagriha to placate the Buddha.[17] When the delegation reached the King of Magadha, he did not give assurance to them for the Buddha's arrival in Vaiśālī as *vassavāsa* was going on and the Tathāgata had to spend his rainy season in the Venuvana, Rājagriha. When the Buddha heard about the problem of the Lichchavis, he, along with other monks, consented to visit them. Bimbisāra made a special arrangement for the Buddha's visit to Vaiśālī. He arranged two large boats for the Buddha, Ānanda, and the other 500 monks accompanying them. Bimbisāra also comes to see him off on the bank of River Gaṅgā. On the other side, the Lichchavi King came to receive the Buddha with complete state honour by going deep into the River Gaṅgā. It is mentioned that the kings, spirits, nāgas, and devas residing in different realms also came to receive the Buddha.[18] It was a unique practice adopted by the Buddha to serve the community with the help of the state. There is parallel evidence in the Later Vedic age and the period of the Buddha that kingdoms/kings took the initiative to resolve crises or the avoidance of any problems linked to their subjects. The foremost evidence is of King Janaka of Mithilā. He used to plough land as part of the Vedic ritual to attain prosperity and avoid drought in his kingdom. On one occasion, his beloved daughter Sītā was miraculously born.[19] It symbolises the importance of agriculture, which led to prosperity and health. The agricultural produce was so vital that it was precious like a king's beloved daughter. Sītā became a symbol of farm produce and a goddess of agriculture. The *Arthaśāstra* mentions the *Sītādhyakṣa* as head of the Department of Agriculture.[20] The second example can be taken from the Śākyans of Kapilvastu. Siddhārtha, while attending an annual ploughing festival led by his father King Suddodhana, entered into meditation and realised the first *jhāna*.[21] The *Ratana Sutta* shows that Buddha became the universal authority by developing a mechanism to eradicate even the most arduous societal problems. As a Dhamma Chakravarti, he led the mission, and the rest of the paraphernalia including the kings and kingdoms rallied behind him.

Contrary to the mechanism and thought process adopted by the Later Vedic Kings where the earth was accepted as the most essential element for the survival and growth of human civilisation, the Ikṣvākus started giving more importance to the sacredness of the water. This notion is well exhibited in the story of the coming of River Gaṅgā from heaven to earth. The Ikṣvākus did an undaunting task to take Gaṅgā to earth. Many left their royal throne and became ascetic to propitiate the gods. King Bhagirath became successful with the help of Lord Brahmā and Śiva to take Gaṅgā to earth.[22] It is deplorable that the Ikṣvākus did such an arduous task to bring Gaṅgā to earth for facilitating *mokṣa* to the 60,000 sons of Sāgara who became a heap of ash after the curse of sage Kapila. The Ikṣvākus were told to be the most benevolent kings in the history of the early civilisation of India and their intention to take the different channels of River Gaṅgā in the mainland was given a mythological attire. The King Viśāla who founded Vaiśālī belonged to the Ikṣvāku lineage and had some social linkages with the ancestors of Rāma. Still, the people of Magadha believe that Rāma, Lakṣmana, and sage Viśvamitra on the way to Mithilā to take part in the *svayaṁvara* of Sītā, visited Vaiśālī to see the kingdom founded by their ancestors. The Ikṣvākus led a paradigm shift where *apo* (water) received more importance than *pāthvi* (earth). Water became the most desirable thing with the rise of urbanisation and growing agrarian activities. Gaṅgā flowing in the mainland of the Indian subcontinent became the lifeline of the newly growing and ever-expanding settlements of the Āryans. They realised the importance of the water of Gaṅgā and paid utmost reverence to her. Gaṅgā or her tributaries became the main thrust of life and economy in the post-Harappan period in the northern part of the Indian subcontinent. The Buddha, the greatest genius of his time, understood the importance of water and extended the scope of its sacredness. *Apo* (water) was considered the most precious among the four significant elements. The Buddha expanded the horizon of the deification of water by adding several subsidiary elements like solemnity with rivers, ponds, and lakes, worshipping of *nāgas* who are considered as bestowers of rain, etc. He also accepted the value of water's regenerative and protective powers and made it an eternal part of Buddhism's social-cultural and ethical domain. The water ceremony embedded in the *Ratana Sutta* began this tradition.

The *Aṭṭhakathās* inform that after he arrived in Vaiśālī, the Buddha preached the *Ratana Sutta* to Ānanda and also instructed him the method of recitation. After learning the whole procedure Ānanda and the Lichchavi princes roamed in every part of the city, sprinkled the sacred water from the Buddha's alms bowl and recited the *Ratana Sutta*. Because of continuous recitation and the sprinkling of water, all kinds of evil spirits fled from Vaiśālī.[23] The Pali texts also inform that after the conclusion of the ceremony, the Buddha preached this *Sutta* to all the citizens of Vaiśālī and the devas of the two realms headed by Sakka for seven consecutive days in the royal assembly hall of the Lichchavis.[24] The people were cured of physical as well as mental defilements. The Buddha created a kind of dissemination agency: the monks to recite the *Sutta* and perform the ceremony. He invented the *Sutta* so that his disciples could recite it on his behalf to benefit all sentient beings, including humans and devas. The recital of the *Sutta* and sprinkling of water has two primary purposes, first to create awareness about hygiene and cleanliness, in which water plays a significant role. The understanding of the social phenomenon of hygiene and cleanliness was linked to the Buddhist ethos and the divine aura of the Buddha. It created a mystical bond among the people to use water to clean their habitation. The ongoing crisis of Vaiśālī was connected to the evil spirits causing an epidemic and only the Buddha's cosmopolitan outlook could remove this menace with his divine power. The recital of the *Sutta* compelled the evil forces to go away. It also introduced a load of religious injections that the people have to respect all kinds of sentient beings and spirits and should regularly pay homage to them.

The *Sutta* also discussed the greatness of the Buddha and his supramundane status. The Buddha attained *Nibbāna* and earned ideals of non-attachment. His Dhamma ideals are unique and unsurpassable. It cannot be compared with other types of virtues. The Buddha realised the truth for the happiness of the whole world and extolled the Noble Eightfold Path, calling it the path that unfailingly brings sublime truth and happiness.[25] The *Sutta* also eulogises the role of the Saṅgha for the welfare of the people. It is subtly endorsed in the ceremony of the *Ratana Sutta*. The Buddha did not personally go to sprinkle water and chant the *Sutta*, but Ānanda took his alms bowl to represent the Buddha. It shows how the Buddha fixed the role of the Saṅgha as a disseminating agency for the propagation of the Dhamma among the lay people. As the foundation of the Saṅgha, the monks are in the leading role of chanting and performing ceremonies. Their role, attributes, and acumen were exalted in the *Sutta*. They are worthy disciples of the Buddha and deserve offerings. The Saṅgha is a precious jewel and many of the respected monks attained the *arahanthood*. Their strength is like a post deep-planted in the earth which stands unshaken by the winds blowing from the four quarters; in the same way, the monks can comprehend the path guided by the Buddha and the ethos of the Noble Truths.[26] Through this *Sutta*, the cosmogony of Buddhism was superimposed in every direction from the terrestrial to the celestial world. All beings, whether devas, humans, or non-humans, accept the authority of the Buddha and pay reverence to him. This cosmological aura is created not for divine superimposition but for establishing peace and happiness in the world. The genuine respect of the Buddha will be when all territorial and celestial beings will respect the Saṅgha and Dhamma.[27] The Buddha resolved a mechanism to integrate the terrestrial and celestial realms with his cosmic aura and set in a method of prognostication that helps the people to develop faith in his mechanism and make decisions to protect themselves from evil spirits. The Pali sources mention that the Buddha's influence reigns supreme over 100,000 million universes with the circumference of which the *paritta* is efficacious.[28] He is described as *akūtabhaya* (without any fear).[29] The *Sonadatta Sutta* of the *Dīgha Nikāya* informs that in whatever town, city, or village, the Buddha goes to reside, non-humans cannot harm the people of that region or village (*Gotamo yamiṃ gāme vā nigame vā pativastina tasmiṃ gāme vā nigame amanussa manusse vihithanti*).[30] The Buddha's gesture of *Abhayamudrā* shows his benevolence and cosmic aura. He is always kind and altruistic with his blessings.

Social Paradigms of *Paritta* and Water Ceremony

The performance of the water ceremony with the chanting of the *Ratana Sutta* opens a new chapter in the history of Buddhism. The Buddha was against performing any ritual and showed the mystical powers in the public domain. Water with the chanting of the *Suttas* became a matter of debate. Many scholars suggest that the *parittas* indicate a belief in supernatural forces and the rectification of cause by ritual performances. In the *Dīgha Nikāya*, the term *parittāna/saraparittānam* is described 13 times for protection against defilements and commonly linked to the moral aptitudes and hurdles coming in the path of *arahanthood*.[31] The *Brahmajāla Sutta* says that monks should not perform ceremonies for monetary gains and favours. Some of the ascetics of the other faith earned their livelihood by practising the same. The *Sutta* categorises

> *palmistry, divining by signs, portents, dreams, body marks, mouse-gnawing, fire-oblations, oblation from a ladle, of husks, rice powder, rice grains, ghee or oil, from the mouth or of blood, reading the figure tips, house and garden lore, spell in charms, ghost lore, crow lore, snake lore, poison lore rat lore, bird lore, crow lore, foretelling a person's life span, charms against arrows, knowledge of animal's cries, etc. The Buddha refrains from acts and wrong means of livelihood.*[32]

The *Upali Sutta* of the *Majjhima Nikāya* mentions that a disciple of Mahāvīra requests him not to send the lay disciples to the Buddha as he is a *māyāviṅ* (magician) and through his *iddhi* he converts the lay people loyal to other sects to Buddhism.[33] The Buddha is told a magician because he developed enormous *iddhi* through his meditation. Because of his charismatic personality, he won over many followers from his opponents. It also reveals that his contemporaries like Mahāvīra and Makkhaliputta Gośāla did not endure the superhuman qualities of the Buddha.[34] Some contexts are also discussed for Moggalāna who was known for his *iddhi-pātihāriya*.[35] The *Dhammapada Aṭṭhakathā* says that he could transform Mount Sineru into the form of a kidney bean.[36] Once the Buddha and his monks did not receive alms in the Verañjā region, Moggalāna requested the Buddha to turn the earth, so that the kernel of the earth lying under the surface might be a source of food. He also consented to open a difficult pass in the region of Uttara Kuru to facilitate safe passage for monks going for alms. The Buddha refused to accept his advice.[37] The *Cūlatanhasankhāya Sutta* says that Moggalāna once visited Sakka to know if he appreciated the sublimity of the teachings of the Buddha but found him full of pride. He shook Sakka's palace, Vejayanta. Sakka was frightened and submitted to the sojourn of the Dhamma.[38] Moggalāna also went to the Brahma world to subdue the arrogance of Baka Brahmā and questioned him in the Sudhammā Hall in the Brahma world. He later confessed his erroneous views and submitted to the Buddha.[39] The *Aṭṭhakathā* says that Moggalāna showed the power *of iddhi* to subjugate the Nāga Nandopananda.[40] The Buddha on one occasion exhorts Piṅdola Bharadwaja for using his *iddhi* to fly and take an alms bowl made of sandalwood placed upon a ground where criminals getting death sentences were used to be hanged.[41] The *Kevatta Sutta* informs that Kevatta of Nālandā requested the Buddha to allow the monks to use *iddhi* (*iddhi-pātihāriya*) or the superhuman attributes (*uttari-manus-dhamma*). It will help to convert the lay people to embrace Buddhism. The Buddha was unwilling to allow the monks to perform any miracles, but later on permitted certain conditions. There are three types of miracles that a monk can realise by attaining the higher stage of knowledge (*abhiňňa*). These three are the superhuman powers (*iddhi-pātihāriya*), the miraculous power of telepathy (*adeśana-pātihāriya*), and the miraculous display of instruction in dhamma (*anuśāsani-pātihāriya*). Despite their ability to perform miracles, the Buddha always preached to the monks to escape from performing them and endorsed that the Dhamma was the true miracle.[42] Luis Gomez suggests that the Buddha declined Kevatta's request. He took a rational view of the miracles and allowed only moderate intervention in case of any vital necessity.[43] The *Mahāvastu* has the same list of miracles with the addition of *dharma-deśana-pārtihāriya*, that is the teaching of the Dhamma.[44] Vasubandhu says that the teaching of the Dhamma is the best. There is a limitation of the effect of charms and spells. The *Gandhārī-vidyā* gives a person power to flight and the *Ikṣaňika-vidyā* capacitates a person to read the minds of others. However, the monks can share the true knowledge delivered by the Buddha, and miracles can be used in moderation. The Dhamma is the best agency to disseminate knowledge for the holistic well-being of the society and the monastic tradition. The teachings of the Buddha facilitate the *Arahantahood* for the monks, the most precious possession for them, and it is the best kind of miracle.[45] The miracles must be spiritual and divine. It must be aimed at serving the larger interest of society. It is a magic spell or psychic event if it is not developed or performed for welfare causes.[46] It is also believed that the Buddha used *iddhi* power to convert the Kassapa brothers. The *Ādittapariyāya Sutta* informs that the Buddha preached this *Sutta* at Gayasisa after the conversion of the Tebhatikajatila, that is Uruvela Kassapa, Nadī Kassapa, and Gayā Kassapa. At the end of the discourse, the Jatilas and other monks became *arahants*.[47] In holistic terms, it was advisable for the fraternity of the Saṅgha to avoid using the *iddhi* power to show any miracle or to convert laity. He refuted all these beliefs and practices to be performed by the monks to earn material gain and the above practices mentioned in the *Brahmajāla Sutta* were prohibited by him to earn money but some of the practices like protection against snake bite or any type of physical injury were adopted by the Saṅgha and the *Parittas* were developed for them. Many of the social values prevailing in the land of Magadha and Kośala during his time were embraced by the Buddha more logically and rationally. The superstitious rituals and dogmatic customs were vehemently rejected by him, but some rational practices related to health and community services were accommodated into the Buddhist fold.

Lily de Silva says that the Buddha was against performing any rituals, incantations, or magical performances (*tiracchavijja*) by the community of monks and nuns. In the Northern Black Polished Ware (NBPW) (6th BCE) period, the people of the Magadha and other adjoining regions strongly believed in rituals, magics, and miracles. The newly established religion needs some alternative mechanism to eradicate these beliefs. The denouncement and disliking of the rituals could not be a substantial measure to counter these subversive activities. The *paritta* developed for the first time in the *Ratana* and the other *Suttas* were the efficient tools to counter the magical incantations. The Buddha developed it and was not contrary to the rules he endorsed.[48] It is believed that such kinds of practices were borrowed from the local traditions and these were non-Buddhist in origin.[49] Melford Spiro sees a kind of dichotomy between *saṁsāric* and *asaṁsāric* worldviews of Buddhism. The magical incantations and rituals are oriented to worldly actions and do not interest the monastic community. The monks do not prefer it because its metaphysical foundation depends upon myth and it is futile for Buddhists in quest of *Nibbāna*.[50] Donald Lopez says that the monastic tradition developed by the Buddha was oriented towards the *nibbānic* quest and gave less emphasis on the material and worldly life.[51] However, the Saṅgha does not consist of only the monastic community. Lay persons were an important part of the Saṅgha and Buddhism did not deny them the right to material prosperity and worldly comfort. If it was the duty of the lay persons to take care of the monks and monasteries, it was also the duty of the monastic community to take care of the interests of lay people. The Dhamma-dāna was an essential part of a monk's duty, and for the safety, progress, good health, and prosperity of the laity, the *parittas* and water ceremonies were developed. The community of monks could not train them in meditation or the subtle teachings of the Buddha. Certain mechanisms had to be evolved for the common people to provide them mental and physical safety. The *paritta* was one such mechanism developed by the Buddha. Since the beginning, either the monks recited for themselves as seen in the Khaṅda-paritta or it was recited for the laity as in Vaiśālī. Its scope was extended to the Buddhist community including monks, nuns, and lay people. The methodology of meditation was invented by the Buddha and well expounded in the Buddhist canonical literature. It is considered as the balanced path of Buddhism which focuses on the *saṁsāric* truth and the nature of human existence. Some of the spiritual practices and ceremonies were borrowed from other traditions, and many were not aligned with the teachings of the Buddha. Despite that, these were very popular and became an important segment in Buddhism's material and social fabric. The Buddha distinguishes between the rightful ceremonies and the supernatural conduct, which were denounced by the Buddha, even if some of the senior monks performed it. Reynold says that the role of magic and Buddhist virtues coincide when there is some major upheaval in society in the form of physical injury, epidemic, sickness, and natural disaster. On certain occasions, the harmful spirits try to destroy the nascent fabric of society because they do not like the unnecessary intrusion of humans in their abodes like ponds, lakes, oceans, or hills. When their wrath led to disaster, an urgent mechanism was required to placate and tame them. In these cases, Buddhism took the help of other traditions and embraced some rituals developed as *parittas*.[52] Craig Reynold says that the six years of wandering of Siddhārtha and the consequent struggle between good and evil gave him wide experience to cope with the social evils. The earth and water played the decisive role in his quest for *nibbāna*. In his continuous confrontation with Māra, Mother Earth always comes to help Siddhārtha. When he requests her to witness his strenuous effort to attain *nibbāna*, she comes out and causes a great deluge that wipes away Māra's evil forces.[53] Water played a decisive role in transforming Siddhārtha into the Buddha. The crossing of River Anoma began his quest of *nibbāna*, crossing the Gaṅgā after leaving Alara Kalāma developed his resolve of steadfastness, River Niranjaṅā paved the way for his Buddhahood, and finally, the water carried by Mother Earth swept away the evil elements. The Water Ceremony of the *Ratana Sutta* symbolises the sweeping away of all evil spirits causing havoc to the people of Vaiśālī. Anne Blackburn says that the development of *paritta* was not an accommodative device but an emphatic social mechanism where the masses believed in a mystical instrument to protect themselves from all kinds of dangers. In this ceremony, the monastic community develops a strong social connection with the laypeople.[54] It can be performed for sacred as well as secular causes in a very simple manner or with much show and pomp. Any kind of sacred boundary or time limitation is not defined for its performance. The dissemination agency has not changed since its inception in the age of the Buddha. The monks can recite the *Sutta* and perform the ceremony to avoid danger and inculcate happiness. It has a very simple kind of ceremony as first discussed in the *Ratana Sutta* and other Pāli *Suttas* but with time, the complex structure of the ceremony emerged as visible in the Theravāda countries like Sri Lanka, Thailand,

and Myanmar.[55] Richard Gombrich says that the use of water and thread is a mandatory part of *paritta* and the importance of it was deliberate in the early Buddhist literature.[56] The use of sacred water and thread has been known as a religious practice since the pre-Buddhist period. The Śākyan believed in different types of water ceremonies. The Buddha might have picked a few of them from the earlier traditions. The *Jātaka* mentions that a prince was protected by the *paritta* and sacred thread that was recited by a Paccekabuddha.[57] S.J Tambiah says that the notion of *paritta* might be ethical or exorcist and the merit earned is like a life-affirming nature. This performance of such acts was done by the monks who inculcated the power of the Buddha because of their renunciation and attainment of *arahanthood*.[58] The efficacy of the *paritta* is ensured by bringing together the three jewels the Buddha, the Dhamma, and the Saṅgha by the recitation of the *Suttas*. The power of three jewels comes together in the form of the *paritta* to protect the interest of the people.[59] The recitation of *paritta* in the Pali language is considered efficacious because it was the word of the Buddha and the blessings are coming from his mouth. The *Sutta* recited in the ceremony belongs to the canonical texts and the monks performing such activities also seek his monastic lineage directly from the Buddha.[60] The recital of the *Ratana Sutta* shows that the *parittas* seek power from different sources but all the sources come from the Buddha. The Buddha is portrayed in the paradigmatic texts as a mentor of the cosmos holding the whole ontological space. He was not divinely grace with such potencies but earned the cosmic power during his six years of wanderings that transformed him from Siddhārtha to the Buddha. The *parittas* exhibit a unique phenomenon that the Buddha is the only cosmic force and the rest of the sources and energies are borrowed from him. His *nibbānic* attributes compel the other including the divines to seek power from him whether physically or symbolically. The *Ratana Sutta* shows that he is the ultimate source of the truth and deliverer of the *metta*. His attribute of loving-kindness will calm down those who are inclined to harm others. The protective verses delivered by the Buddha ensure that monks and lay people must develop compassion and loving-kindness for those who tend to harm.[61] There is nothing fictitious about the *paritta*. It ensures love, affection, compassion, and good virtues. It inculcates positive energy in the mind and body of the person who is affected or who is harming. The sound of chanting sends positive rays of energy and exudes confidence among the participants that the Buddha as guardian of the cosmos will protect them.[62] Because the *paritta* was directly given by the Buddha, the performer or reciter, that is monks, and listeners, that is the people, must participate in the ceremony with full discipline and calmness. Only the sincerest attention can bring the most positive results.

The modes of narration and recitation also decide the effective way of communication. Language influences the way of thinking and the level of acceptance of particular thought processes by the members of society. This ushers in and links the individuals and the community to transmit their desires in a set manner and to establish procedures, means of communication, limitations, and linkages. The language seeks its strength from the content, intentions, purpose, and way of performance by the reciters.[63] In the Buddhist ceremonies related to well-being, it is common practice to recite the *suttas* loudly, because the syllables and the content mentioned in the *suttas* are meant to be heard by the followers and the natural forces of the surroundings. This is one of the ways of communication and decides the efficacy of the purpose of the ceremony.[64] Many texts and references mention the canonical texts and the *Aṭṭhakathās* are not examined properly as the scholars dealing with the different disciplines of Buddha are generally concentrated upon the monastic traditions. Because of this shortcoming, some of the studies lack the true face and historicity of particular traditions of the early phase of Buddhism or in contemporary Buddhism.[65] Piyadassi Thera says that the *parittas* were not only used for prophylactic purposes but their scope was extended in the monastic training. The young novices were trained in the *paritta* literature and through this knowledge they trained themselves in the conduction of the ceremonies, the exact way of the recitation of the suttas, and also memorising the important suttas, once told by the Buddha.[66] The *parittas* became a training manual for young novices. The selected anthology of the Buddha's discourses was compiled by the ancient monks and is now accepted as the manuals for training the *sramaneras*. The purpose of this kind of training is that many of the monks are not able to comprehend the large *Suttas*, and must learn at least the basic curriculum on protection. By doing this, they will be accentuated with the essential teachings of the Buddha. In the learning process, the novices learn to recite the text as described in the manual and learn the way of recitation, kind of frequency of loudness during the process, melody during the recitation, the meaning as well as the contexts of the syllables mentioned in the Pali *Suttas*. They also ponder over the antiquity and historicity of the texts, contexts, and purpose. It also gives the idea of how these *parittas* became so efficacious. This training allows them to achieve perfection in the performance of the parittas and links them with the larger world of the laity in and outside the ritual arena.[67] This unique mechanism was binding the monastic community and the lay world. Both have different aspirations, but their cosmic authority is familiar: the Buddha. Since the beginning, the *parittas* have always been recited orally with a definite type of melody and sound. The *Sutta* is developed and compiled in the textual form, but the phonetic characteristics of the *Sutta* are an essential feature of the *parittas*. The mode of recitation and phonetic expression guide the reciter's mind to set the tone with the Buddha's discourses. It seems that some recital structure was developed in the form of the *Gāthās* and *Padas* but the Buddha intentionally discouraged taking inspiration from musical notes. It was a significant departure from the Vedic tradition where music was important to reciting the Vedic hymns.[68] The Buddha also prohibited the use of musical instruments in partita recitals; only water and thread were essential elements of the *paritta* ceremonies. The Buddha warns the monks not to be infatuated with the musical notes. The later compilers of the *Suttas* took care of this instruction and carefully avoided lyrical continuity in the *sutta*. Sometimes they added *mātrā* to disrupt the regular flow of recitation.[69] It is noticeable that the Buddhist reciting tradition of the *parittas* was developed with great care. The sound of the chanting not only removes doubts from the minds of the laity but also acts as a carrier to carry forth the truth and well-being.[70] The pattern of recitation, water sprinkling, and thread binding are focused on the well-being and welfare of the individual, society, and world at large.

Types of *Parittas*

The *paritta* consisting of 'blessed water' developed by the Buddha in the *Ratana Sutta* was the first attempt in Buddhism to safeguard the familiar people against natural perils, epidemics, weapons, health hazards, and other harmful events. In this mechanism, the harming spirits were compelled to leave the habitation area and accept the suzerainty of the Buddha. The mode of recitation of the *sutta* and water sprinkling are the prime factors for the success rather than the absorption and imbibition of the ideas and process by the laity.[71] The recitation process normalises and neutralises the power of the spirits they seek from their respective landscape and the energies of the earth. Their injurious energies are channelled or harnessed helpfully by correctly reciting the *sutta* to protect human beings. The sprinkling of the water soothes the negative energy of the earth, created by the evil spirits. With the rising popularity of Buddhism and the consequent expansion of religion in different parts of the world, the scope of the *parittas* was also extended. Several modes of recitation were adopted for various purposes. It was not only for the cure of sickness or epidemics but also for other issues.[72] One of the earliest notions of *paritta* is found in the *Cullavagga* and the *Aṅguttara Nikāya* when a young monk died of a snake bite. The *Ahina Sutta* of *Aṅguttara Nikāya* mentions the *parittas* to provide safety to the monks which prevents the monk from being bitten by a *nāga*.[73] The Buddha preached compassion for the snake and advised the monks to save themselves from this kind of unfortunate incident.[74] The *Khanda Paritta* is one of the oldest *paritta* texts in which the Buddha developed loving-kindness towards the nagas. Even if the monk died of a snake bite. The *paritta* informs that with the divine powers of the Buddha, loving-kindness will be developed among all sentient beings.

> *Sabbe sattā sabbe pāṇa, sabbe bhūta ca kevalā;*
> *Sabbe bhadrāni passantu, mā kañci pāpam āgamī.*
> (All beings, all creatures, everyone- May all carry good fortune and not come to any harm.)
> *Appamāṇo buddha, appamāṇo dhammo;*
> *Appamāṇo saṅgho, pamāṇavantāni sarisapāni.*
> (The strength of the Buddha is unaccountable, the Dhamma is unmeasurable. The Saṅgha is unmeasurable but the crawling creatures are measurable.)[75]

The *Karaniya Metta Sutta* also discusses this.[76] The *Maṅgala Sutta*, like a previous *Sutta*, is small, having only 12 verses, but is believed by the Theravada Buddhists as one of the most powerful *Suttas* for the welfare of the universe. The recitation of this *Sutta* removes unfortunate incidents and brings auspiciousness to personal and public life. The anthropological worldview of this *Sutta* envisions the challenges and finds the solutions in the diverse Indian conditions where Buddhism was born.[77] The *Maṅgala Sutta* is devoid of metaphysical, spiritual, or ritual obligations. The Buddha enumerates different blessings in these suttas as modes of blessings, auspiciousness, and kindness. These righteous ways could be permeated into and accepted by all sections of society. The *Mora Jātaka* informs the remarkable effect of *paritta* chanting. The peacock who used to chant the *sutta* in the morning and the evening lived for hundreds of years without any fear or danger. When he obliterated the recitation because of a fatal attraction towards a peahen, he fell into a hunter's net trap.[78] The example of the *Mora Jātaka* shows that the *Sutta* recitation was not the monks' exclusive domain. The laity and other beings can also recite for their safety and wellbeing. The *Supparaka Jātaka* mentions an episode of a shipwreck during a hail storm. The Bodhisattva, travelling on the same ship, performed the *paritta* ceremony. He bathed with the scented water, put on a new robe, started reciting the sutta, and drained out the ship's water with both hands for the safety of the people. The boat was saved and reached its destination safely.[79] Faxian mentions a similar situation when he was returning to China from India. When he boarded a ship in Jāvā, a sudden storm appeared after two days of journey and the ship developed a leak. Faxian recited the *maṅtra* to worship Avalokiteśavara and prayed for safety. The storm subsided and the ship reached Sumatra.[80] In the *Milindapañha*, a long list of *parittas* is provided for different purposes. King Milinda questioned the efficacy of the *parittas* and asked several questions from Nāgasena about the effectiveness of the *paritta*. Can it save the person from death? How can it be effective and under which conditions? Nāgasena replied:

> *Not in the sky, not in the ocean's midst,*
> *Not in the most secluded mountain cleft,*
> *Not in the whole wide world is found the spot*
> *Where standing one could escape the snare of death.*[81]

The Buddha adopted the *parittas* for the general well-being of the people. The *Ratana Sutta*, the *Khaṅda Paritta*, the *Mora Paritta*, the *Dhajagga Paritta*, the *Aṅgulimāla Pariita*, etc., were developed on different occasions. Nāgasena endorsed the Buddhist view that not in any circumstance, a person can escape death. Still, *paritta* saves the person from unnatural happenings and assists them in living their whole life in peace and happiness. It applies to those who have some life span left. If age is mature enough, the *paritta* will be like pouring thousands of pots of water on dead wood that will not be sprouted again. It protects the people from misfortune. For example, recitation of *paritta* will save him from the snake bite. It is like medicine that cures the person if given promptly. Only in case of obstruction of lousy karma, sin, and disbelief in the Buddhist practice will the paritta not work. Performing asseveration or solemn declaration of the truth (*saccakriyā*) is mandatory for the sacrament's efficacy. When an act of asserveration of truth is performed, there is a positive impact, like there will be good rain or the fire will be subsided, etc.[82] Aṅgulimāla as a monk pioneered developing the *paritta* for the safety of pregnant women. By reciting this, he could ease a woman's labour pain. His recitation was known as *saccakiriyā* or act of truth (*yato aham sabbaññutabuddhassa ariyassa ariyāya jātiyā jāto*). Later it became developed as a *paritta* to ward off all dangers. The water that washed the stone on which he sat in the woman's house came to be observed as a panacea.[83] The *Saṁyutta Nikāya* mentions three *Suttas* in one place describing a sick person who recites the *bhojaṅgas* (seven factors of enlightenment) to get rid of his disease. These verses are known as the *Gilana Sutta*.[84] The *Vinaya-Vinicchaya*, text of the 4th–5th century CE, written before the *Aṭṭhakathās* in Sri Lanka shows the importance of purified water for sprinkling and the sacred thread to be tied. It informs that a sacred thread tied to the lay community must be sanctified with the purified water of the *paritta* ceremony. After that, the thread will become more efficacious.[85] Later, special chanting types dedicated to particular monks were also developed as the *paritta*. John S. Strong finds that the medallion of Upagupta is very popular in Thailand and Myanmar. Bhikkhu Upagupta is always associated with Emperor Aśoka and it is said that he accompanied Aśoka in his Dhammāyatrās to the different Buddhist pilgrimages. In Thailand, it is popularly known as Phra Uppakhut amulets depicting monk Upagupta looking at the sun and holding his alms bowl, and on the other side, a magical charm is shown. Such kinds of amulets protect the person from injury and accidents. The amulets function at the personal level and the images of Upagupta act at the community level, like in the festivals, social functions, etc.[86] The *Mahāvaṁsa* does not mention any *paritta* performed by the kings in Sri Lanka but informs that when Prince Vijaya started his journey to Sri Lanka, on the instruction of Sakka, he was protected by the sacred thread by God Uppalavana. Because of such protective measures, the yakṣas could not harm the prince.[87] Lily de Silva argues that the *paritta* ceremony to protect the prince's journey was added later in the Vijaya legends. Even in the grand ceremony of consecration of the Mahāthūpa by King Duttagamini, *paritta* has not been recited.[88] The first ceremony of the paritta was mentioned in the *Culavaṁsa* when King Upatissa performed the same kind of water sprinkling ceremony as Ānanda performed in Vaiśālī.[89] It is believed in Sri Lanka that at the time of his *mahāparnibbāna*, the Buddha instructed the Sakka to protect the interest of the people of Sri Lanka. Sakka informed Viṣnu to go to Sri Lanka to perform the water ceremony. He sprinkled water and tied a sacred thread on essential places to protect the people of Śri Lanka.[90]

In the Mahāyāna, a new kind of genre, that is, *rakṣā* is developed, dealing with mundane desires like seeking good health, safety from perils, calamities, etc. These *mantras* are not found in the canonical literature but were developed by the followers of Mahāyāna. One of the essential characteristics of these *mantras/dhārinis* is that their syllables might be intelligible and unintelligible. It can be recited, put in the stupas as a relic, or used as a banner in peace marches, the battlefield, etc.[91] In the *Mahāyānsangrha*, the Buddha was worshipped in a very personal way and all the prayers are dedicated to him to protect the people from problems. It is presumed that the omniscience of the Buddha will always be protective and benevolent.[92] Peter Skilling has extensively discussed the types and nature of *parittas* and related water ceremonies, especially in the Mahāyāna and Vajrayāna *Sūtras*. He calls it *Rakṣā Sūtras* like the *parittas* of the Theravāda tradition. The only difference between the two is that the *parittas* are mentioned only in the Pāli literature but the *Rakṣā* (*rakkhā*) *Sūtras* are mentioned in both Pali and Sanskrit traditions. These two kinds of literature exist and are practised across all Buddhist traditions. The purpose of this invocation is to ensure the safety of human beings from natural disasters, calamities, snake bites, floods, maligning spirits, etc.[93] One class of literature of the Mahāyāna is the *Mahāsūtras*. These are the eighteen *mahāsūtras* listed in the *Vinayas* of *Sarvāstivādins* and Mūlsarvāstivādin. In these *Sūtras*, several *Rakṣā Sūtras* are accepted at par with the *Ratana Sutta*. The *Mahāsahasrapramardonī Pancarakṣā* is also considered equal to the *Ratana Sutta*.[94] Peter Skilling says standard phraseology is developed in the Pali and the Sanskrit Rakṣa Sūtras. Here the *parittas* and *saccakriyā* are intermingled. In the Mūlsarvāstivādin literature *Sardulakarnāvadāna*, the text is

> *etena satyavakyena svastu anandaya bhksave; anena satyena satyavakyena* in the *Pratihāriya Sūtra*; *tena me satyavakyena* in the *Upāsanā Sūtra*; *etena saccena Savatthi hotu* in the *Ratana Sutta* and the other *Aṭṭthakathās*. In the *Ratana Sutta* of the Lokottaravādins the text is '*etena satyena suvasti hotu*'. In the *Mahāsaharsapramardanī* verses of the *Ratna Sutta*, the verse is '*etena satyena ihastu svasti*'

and in the *Prajñāparamitā* '*anena* '*satyena satyavacanena*'.[95] The different kinds of water ceremonies for protective measures have been devised and performed with other types of syllables, *mantras*, and *dhārinis*. With such kind of expansion of *paritta* ritual, the scope of the deities was also extended. Now, not only the historical Buddha but also the number of the Buddhas, Bodhisattvas, and female deities were included among the pantheon dedicated to the protective ceremonies. The *Suvarnaprabhāsa Sūtra* (*The Sutra of the Golden Light*), probably composed in the 1st century CE in Indian monasteries, found its way into many countries. In this ceremony, Goddess Sarasvatī is invoked and *dhārinis* in her praise is recited. The water for the ritual bathing was mixed with the medicinal plants. The river goddess informs that when she is invoked with full rituals, she appears in person accompanied by the pantheon of gods and goddesses to eradicate diseases, sickness, droughts, fear of violence, etc. She will also protect the people from bad dreams, evil spirits, demons, etc.[96] The *Suvarnaprabhaottama Sūtra* was translated into Chinese as early as the 7th century. It gives a fair idea about the Indian system of medicine, the classification of diseases, and rituals linked to the remedies and diseases.[97] It is supposed to be first composed in the Sanskrit language and by the 5th century, its different versions were known in Sogdiana, Xixia, Mongolia, Uighur, Khotan, Nepal, and Tibet. It reached Japan in the 7th century CE. Its earliest Chinese version was written by an Indian monk, Dharmaksema, in 414–421 CE, and its complete account was translated by I-Ching in 703 CE.[98] Despite the emergence of different kinds of literature for protection, their roots are the same. Peter Skilling says,

> *The Paritta of Theravādins, the Mahāsūtras, rakṣās, and mantras of the Mūlsarvāstivādins, and the svastigāthā, rakṣās, and mantras of these and other schools of both the Śrāvaka and Mahāyāna were not independent or isolated development. The chanting of certain auspicious verses or texts of protection against disease and malignant spirits and for the promotion of welfare was no doubt a "pan-nikāya" practice common to all branches of saṃgha from an early date, indeed, on the internal evidence of texts like the Dhvajāgra and Ātānātika Sūtras, the practice should predate the early schisms. The two Sūtras just referred to are parittas and Mahāsūtras; the Ratana Sutta is a paritta, a svastigāthā, and a key element of Pañcarakṣā texts. In some schools or communities, the practice of rakṣā developed further with the use of the mantras and vidyās, by the beginning of the Common Era at the very latest.*[99]

The Buddha offered *paritta* and the notion of holy water for the welfare and compassion of all living beings. The inherent idea was to provide a safety mechanism to monks, nuns, and common people at the physical as well as psychological levels. Its scope and varieties were extended as a monastic system started growing and with the emergence of different sects and traditions, *Mahāsūtras, rakṣās, dhārinis*, and *svastigāthās* were also developed and it became more sacramental in nature than the simple ceremonial procedure of the early period. The mode of recitation and ceremonies may vary but the core value, that is '*etena saccena suvatthi hotu*' (By this truth, may there be well-being!) remained the same in all.

Water Ceremonies in Vajrayāna Traditions

The water ceremony with lighting the butter lamp is essential in Vajrayāna tradition covering the wide land of the whole Indian Himalayan region, Bhutan, and the Tibetan plateau. Etymologically it is known as Yon-Chab. The word 'Yon' means offerings and 'Chab' denotes water. It is not easy to trace its origin among the Buddhist cultures of the Himalayan Hills. It seems that it was a version of water ceremonies developed in the heartland of early Buddhism and carried forward by the Buddhist *ācāryas* who visited Tibet and the Ladakh region from the land of Magadha. Vajrayāna believers in Tibet and Ladakh accept that it was introduced by the great Ācārya Atisa that led to the worshipping of the Buddha and the great *ācāryas*. It is a popular notion to offer seven bowls of water, which represents the seven limbs of prayers – paying homage, giving an offering, repentance of wrongdoings if any, rejoicing in goodness, praying to the Buddha for his benevolence, inviting the Buddhas to teach, and dedicating the merit.[100] For it, the followers of the Vajrayāna offer seven bowls.

1 The first bowl of water (*argham̐*) is drinking water, which is poured on all the Buddhas and the Bodhisattvas with proper invocation. It represents the cosmos of all sentient beings.
2 The second bowl of water (*padyam̐*) is used to wash the feet of the deity with flowers, rose petals, and an offering of butter lamps. It symbolises the purification of unwanted desires and actions.
3 The third one is an offering (*pushpe*) of flowers which symbolises a generous and conscious mind.
4 The fourth bowl represents offerings of incense (*dhupe*) mixed with rice. It symbolises morality and good virtues.
5 A bowl of butter lamp (*aloke*) is offered to dispel the ignorance.
6 A bowl of perfumed water (*gandhe*) mixed with rice symbolises the joyous effort.
7 A bowl of food (*naivedya*) as oblation.

Though it is a popular culture that a devotee should offer seven bowls of water, laities believe it is not obligatory to offer seven bowls. A devotee can offer a smaller number of bowls. The offering bowls may be gold, silver, copper, bronze, or earthenware. There must be eight qualities of water: calm, flavoursome, light, soft, clear, odourless, and not troubling the stomach and the throat when drunk. It removes anger and ignorance and produces good karma and virtues. Sometimes, the *dharanī* related to it is also recited by the Lāmās. These offerings bestow honour, material prosperity, knowledge, and blessings of *ācāryas* and the Buddhas (Figure 3.2).

Padmasambhava, the great Ācārya of Nālandā came to Bhutan in the middle of the 8th century CE. Here, he is known as Padmasambhava, or Guru Rinpoche, and is considered a real architect of the Buddhist world of Bhutan. He is often believed as the second Buddha and professed as the personification of enlightened wisdom, compassion, and bliss. Most Bhutanese religious festivals, ceremonies, and prayers are dedicated to his worship. His life and miracles are known through *kathang* hagiographies, written over centuries in the different monasteries of Bhutan. It is believed that in the 8th century CE, Guru Rinpoche buried some Buddhist ritual manuals and other treasures in

Figure 3.2 Bodhisattva with seven bowls, Leh, Courtesy: Stanzin Mingur.

the Burning Lake of Bhutan and it was rediscovered in the 15th century CE by monk Pema Lingpa. The ritual manual of the water ceremony *lhasang* and *trusol* was part of that treasure. *Trusol* is the ablution, and the *lhasang* represents offering ablution with a bathing ceremony for purifying defilements. Perfumed water is put in a special bathhouse. It symbolises the divine presence of the Buddhas and the Bodhisattvas. They are invited by chanting of the *maṅtras* and *dhāranis* to accept the offerings given during the bath ceremony. After the bathing ceremony, the deities were cleaned with soft towels and adorned with new robes. All kinds of physical illnesses, negative vibes, and defilements are believed to be cleansed. In another form of this ceremony, the chanting of the appropriate *maṅtras* reflects the visualisation of the Buddhas melting into sublime liquid flowing upon a person to wash his defilements and bless him with health, longevity, and happiness. In Bhutan, this kind of ceremony is performed in Buddhist households when a family member falls sick or goes through some problem. The Lāmā of the region gives '*tru*' for visualisation of the divinities and pours water on that person's head. It is understood that the person will impart the entire process and his/her problems will be resolved.[101] This ceremony seems to be inspired by the water sprinkler ceremony performed to celebrate the birth of the Buddha as well as from the *Ratana Sutta*. In the early medieval period, bathing ceremonies for the Buddhist deities were very common in Buddhist monasteries across India. I-ching mentions elaborate bathing rituals for images of the Buddha and the Bodhisattva in the monasteries and in the house of laity in the land of Magadha. The discovery of buried treasure is inspired by the Mahāyāna tradition that Nāgārjuna went to *patālaloka* under the deep ocean to discover the great Mahāyāna treatises. The purpose of water ceremonies performed in the Leh–Tibet region and Bhutan is the same, but their different patterns of rituals seem to be oriented from their Ācāryas Atisa and Padamsambhava who developed various types of rituals imbibing the local traditions with the teachings of the Buddha.

Notes

1 Donald K Swearer, 'Consecrating the Buddha', in *Buddhism in Practice*, ed. Donal S. Lopez, Jr., Princeton: Princeton University Press, 1995, pp. 50–51.
2 *Papañcasudānī*, II.918–920.
3 *Buddhacarita*, I.27, *Buddhacarita, or Life of Buddha Venerable Aśvaghoṣa*, tr. E B Cowell, Oxford: The Clarendon Press, 1893.
4 *Majjhima Nikāya*, III, 123.
5 Julia A.B. Hegewald, 'The Lotus Pool: Buddhist Water Centuries in the Kathmandu Valley', *South Asian Studies*, vol. 13, no. 1, 1997, p. 146.
6 *Buddhacarita*, I.40–42.
7 Daniel Boucher, 'Sutra on the Merit of Bathing the Buddha', in *Buddhism in Practice*, ed. Donald S. Lopez, Jr., Princeton: Princeton University Press, 1995, p. 59.
8 Boucher, 'Sutra on the Merit of Bathing the Buddha', p. 62.
9 J. Takakusu, *A Record of the Buddhist Religion as Practiced in India and the Malaya Archipelago (AD 671–695)*, New Delhi: Munshiram Manoharlal, 1998, pp. 148–152.
10 Boucher, 'Sutra on the Merit of Bathing the Buddha', p. 66.
11 Takakusu, *A Record of the Buddhist Religion as Practiced in India and the Malaya Archipelago (AD 671–695)*, p. 45.
12 Donald K. Swearer, *Becoming the Buddha*, New Delhi: Motilal Banarasidas, 2017, pp. 4–5.
13 Swearer, *Becoming the Buddha*, pp. 5–6.
14 Lily de Silva, *Paritta: A Historical and Religious Study of the Buddhist Ceremony for Peace and Prosperity in Sri Lanka*, Colombo: National Museum, 1981, pp. 3–5.
15 de Silva, *Paritta*, p. 139.
16 *Suttanipāta*, VV.222–238.
17 *Mahāvastu*, tr. J.J. Jones, London: Luzac and Company, 1949, vol. 1, pp. 209–211.
18 Bhikkhu Nāṇdassana, *Ratana Sutta, The Buddha's Discourse on the Three Superb Jewels and Their Intrinsic Power- With an Introduction, Translation, and Explanation based on Pali Sources*, Colombo: Sridevi Printers, 2011, pp. 17–18.
19 S. Singaravelu, 'Sītā's Birth and Parentage in the Rāma Story', *Asian Folklore Studies*, vol. 41, no. 2, 1982, pp. 235–236.
20 *Arthaśāstra*, II.24.
21 *Majjhima Nikāya*, I.246–247.
22 Sudipta Sen, *Ganga, The Many Pasts of a River*, Gurgaon: Penguin Random House, 2019, pp. 55–60.
23 *Suttanipāta Commentary*, I.278ff.; *Dhammapada Aṭṭhakathā*, III.436ff.
24 *Dhammapada Aṭṭhakathā*, III.196.
25 *Suttanipāta*, VV.4–5.

26 *Suttanipāta*, VV.7–10.
27 *Suttanipāta*, VV.15–17.
28 Peter Skilling. 'The Raksa Literature of the Sravakayāna', *The Journal of Pali Text Society*, vol. XVI, 1992, pp. 109–182, p. 117.
29 *Theragāthā*, V.510; *Therīgāthā*, V.333.
30 *Dīgha Nikāya*, I.116.
31 Laura D. Frude, *Verbal Magic in Pali Canon, A Study on Protective Chants, Truth-acts, and Vows*, Bristol: University of Bristol, 1991, pp. 21–22. http://research-informationbristol.ac.uk.
32 *Dīgha Nikāya*, I. 9.
33 *Majjhima Nikāya*, I, 375.
34 David V. Fiordalis, 'Miracles in Indian Buddhist Narratives and Doctrine', *Journal of the International Association of Buddhist Studies*, vol. 33, no. 1–2, 2010, p. 384.
35 *Aṅguttara Nikāya*, I, 23.
36 *Dhammapada Aṭṭhakathā*, III.212.
37 *Vinaya Piṭaka*, III.7; *Samantapāsādikā*, I.182f.
38 *Majjhima Nikāya*, I.251ff.
39 *Theragāthā*, V.1198; *Saṁyutta Nikāya*, I.144ff.
40 Theragāthā Commentary, II.188ff.
41 *Vinaya Pitaka*, II.12.
42 *Dīgha Nikāya*, I, 211ff.
43 Luis Gomez, 'The Bodhisattva as Wonder-worker', in *Prajnaparamita and Related System: Studies in Honor of Edward Conze*, ed. Lewis R. Lancaster, Berkely: University of California, pp. 223–225.
44 *Mahāvastu*, I, 237–238, III.137–138.
45 Fiordalis, 'Miracles in Indian Buddhist Narratives and Doctrine', pp. 387–389.
46 Richard Swinburne, *The Concept of Miracle*, New York: St. Martin Press, 1970, pp. 8–9.
47 *Vinaya Pitaka*, I.34–35.
48 de Silva, *Paritta*, pp. 23–24.
49 Richard Gombrich, 'The Buddhist Attitude to Thaumaturgy', in *Bauddhavidyāsudhākaraḥ. Studies in Honour of Heinz Bechert on the Occasion of His 65th Birthday*, ed. Petra Kieffer-Pülz, and Jens-Uwe Hartmann, Swisttal-Odendorf: Indica et Tibetica Verlag, 1997, pp. 165–185.
50 Milford Spiro, *Buddhism and Society: A Great Tradition and Its Burmese Vicissitudes*, Berkeley: University of California, 1970, p. 159.
51 Donald S. Lopez jr., *Buddhism in Practice*, Princeton: Princeton University, 1995, pp. 77–78.
52 Craig J. Reynold, 'Magic and Buddhism', in *The Buddhist World*, ed. John Power, London and New York: Routledge, 2016, pp. 339–341.
53 Craig J. Reynold, 'Power', in *Critical Terms of the Study of Buddhism*, ed. Donald S. Lopez Jr., Chicago: University of Chicago Press, 2005, pp. 213.
54 Anne M. Blackburn, 'Magic in the Monastery: Textual Practice and Monastic Identity in Sri Lanka', *History of Religions*, vol. 38, no. 4, 1999, p. 371.
55 de Silva, *Paritta*, p. 3.
56 Richard Gombrich, *Precept and Practice, Traditional Buddhism in the Rural Highlands of Ceylon*, Oxford: Clarendon Press, 1971, p. 201.
57 *Jātaka* no. 396.
58 S.J. Tambiah, *Buddhism and the Spirt Cults in Northeast Thailand*, Cambridge: Cambridge University Press, 1979, pp. 195–196.
59 Peter Harvey, 'The Dynamics of Paritta Chanting in Southern Buddhism', in *Love Divine: Studies in Bhakti and Devotional Mysticism*. ed. Karl Werner, Surrey: Curzon Press, p. 58.

60 Tambiah, *Buddhism and the Spirit Cult*, pp. 197–198.
61 Eviatar Shulman, 'The Protective Buddha', *Numen*, vol. 66, no. 2/3, 2019, pp. 213, 232–233.
62 V.F. Gunaratna, 'Foreword', in *The Book of Protection*, Piyadassi Thera, Kandy: Buddhist Publication Society, 1999, pp. 11–21.
63 Frude, *Verbal Magic in the Pali Canon*, pp. 23–24.
64 S. J. Tambiah, 'The Magical Power of Words', *Man*, vol. 3, no. 12, 1968, p. 179.
65 Todd Lewis, 'A History of Buddhist Ritual', in *The Buddhist World*, ed. John Power, New York: Routledge, 2016, pp. 318–337.
66 Piyadassi Thera, *The Book of Protection*, Colombo: H.M. Gunasekara Trust, 1975, pp. 5–6.
67 Jeffrey Samuel, 'Text Memorized, Text Performed: A Reconstruction of the Role of Paritta in Sri Lankan Monastic Education', *Journal of the International Association of Buddhist Studies*, vol. 28, no. 2, 2005, pp. 362–363.
68 Paul D. Greene, 'The Dhamma as Sonic Praxis: Paritta Chant in Burmese Theravada Buddhism', *Asian Music*, vol. 35, no. 2, 2004, pp. 44–46.
69 Cyril de Silva Kulatilake, 'Buddhist Chant in Sri Lanka and Its Musical Elements', *Jahrbuchfur Musikalishe Volksund Volkerkunde*, vol. 10, 1982, p. 27.
70 David Kalupahana, *A History of Buddhist Philosophy: Continuities and Discontinuities*, Honolulu: University of Hawaii, 1992, pp. 60–63.
71 Richard Gombrich and Gananath Obeysekera, *Buddhism Transformed: Religious Change in Sri Lanka*, Princeton: Princeton University Press, 1988, pp. 19–20.
72 Donald K. Swearer, *The Buddhist World of Southeast Asia*, Chaing Mai: Silkworm Publishers, 2009, pp. 28–29.
73 *Aṅguttara Nikāya*, II. 72–73.
74 *Vinaya Pitaka*, II,109; *Aṅguttara Nikāya*, II,72.
75 *Aṅguttara Nikāya*, IV.68.
76 *Suttanipāta*, VV.143–152.
77 Charles Hallisey, 'Auspicious Things', in *Buddhism in Practice*, ed. Donald S. Lopez Jr., Princeton: Princeton University Press, 1995, p. 413.
78 *Jātaka*, no. 159.
79 *Jātaka*, no. 463.
80 Samuel Beal, *Travels of Fah-Hian and Sung-Yun from China to India (400 AD & 518 AD)*, Delhi: Low Price Publication, 2005, pp. 166–167, reprint.
81 *Milindapanha*, p. 213, The Question of King Milinda, tr. T.W. Rhys Davids, vol. 1, New Delhi: Low Price Publication, 2006, reprint.
82 *Milindapanha*, p. 120.
83 *Majjhima Nikāya*, I. 103–104; *Papañcasūdanī*, 747f.
84 *Saṁyutta Nikāya*, V.79–81.
85 *Vinaya-Vinicchaya*, VV. 491–492.
86 John S Strong, *The Legend and Cult of Upagupta, Sanskrit Buddhism in North India and Southeast Asia*, Delhi: Motilal Banarasidas, 1994, pp. 277–278.
87 *Mahāvaṃsa*, VII. 914?
88 De Silva, *Paritta*, p.19.
89 *Cullavamsa*, III. 189–195.
90 *Mahāvaṃsa*, VII. 5–8.
91 Peter Skilling, The Rakṣā Literature of Śravakayāna', *Journal of Pali Text Society*, vol. 16, pp. 166–167.
92 Paul J. Griffith, 'A Hymn in Praise to Buddha's Good Qualities', in *Buddhism in Practice*, ed. Donald S. Lopez, Jr., Princeton: Princeton University Press, 2007, p. 44; Paul J. Griffith, Noriaki Hakamaya, John P. Keenan, and Paul L. Swanson, *The Realm of Awakening: A Translation and Study of the Tenth Chapter of Asnga's Mahayanasangraha*, New York: Oxford University Press, 1989, pp. 123–168.

93 Skilling, 'The Raksa Literature of the Sravakayana', p. 110.
94 Skilling, 'The Raksa Literature of the Sravakayana', pp. 131–132.
95 Skilling, 'The Raksa Literature of the Sravakayana', pp. 144–145.
96 R.E Emmirick, *The Sutra of Golden Light, Being a Translation of the Suvarnaprabhasa Sutra*, Oxford: Pali Text Society, 2004, pp. 45–47; Catherine Ludvik, *Recontextualizing the Praise of a Goddess: From Harivamsa to Yijing's Chinese Translation of the Sutras of Golden Light*, Kyoto: School of East Asian Studies, 2006.
97 C. Pierce Salguero, 'On Eliminating Disease: Translation of Medical Chapter from the Chinese Versions of Sutra of Golden Light', *Journal of Indian Medicine*, vol. VI, 2013, pp. 21–22.
98 Emmirick, *The Sutra of Golden Light*, pp. xi–xiii.
99 Skilling, 'The Raksa Literature of the Sravakayana', p. 168.
100 Khenpo Gyaltsen, A Lamp Illuminating the Path to Liberation: An Explanation of Essential Topics for Dharma Students, 2014, pp. 86–92, Lhasey Lotsawa Translations, https://www.lhaseylotsawa.org (The inputs about water ceremony in the Leh-Ladakh region is taken from Dr. Stanzin Mingur, Assistant Professor, Central Institute of Buddhist Studies, Leh (India during personal conversation. He is a practicing Buddhist and a follower of Vajrayāna Buddhism.).
101 Karma Phuntsho, *The History of Bhutan*, Noida: Random House, 2016, pp. 130–140.

4 Nāgas and Aqua-Environment in Buddhism

Water is considered sacred. It is believed that the paramountcy, power, and sacredness are boons of spirits protecting the water and its sources. Their magico-spiritual power lies at the bottom of the sea and is bestowed occasionally to human beings by nāgas. In Sanskrit, the word 'nāga' denotes snake or serpent. They are considered semi-divine creatures, possessing hundreds of palaces, magnificent houses gardens, enormous wealth, living in a magnificent abode, etc. Their habitation *nāgaloka* or *patālaloka* is a nether world, situated beneath the earth in the great ocean, crowned with multiple luxuries. It is also known as *rasātala* or *nirāya*.[1] Both the Brahmanical and Buddhist literature describe the Nether world or Pātālaloka as a prosperous place divided into seven realms. Sometimes it is considered better than Indra's abode. The abundance of richness and material prosperity makes this place Bhogavatī (land of pleasure).[2] The nāgas reside in seas, oceans, lakes, pools, wells, and springs. Their worship always helps to maintain a cordial bond with water, a genesis of life and prosperity. As serpents are always found near water sources, they are considered the guardians of it and the bestower of life, eternity, and positive energy linked to life.[3] In Indic religious traditions, Nāgarājas or the snake deity is linked with rain, fertility, and water. They maintain a spatial relationship with Nature, water reservoirs, lakes, ponds, and rivers.[4] They are worshipped as folk spirits possessing the mysteries and strengths of Nature and the beholder of rain. Because of that, nāgas are considered guardians of the agrarian economy, the bestowers of prosperity, and happiness. They are often represented in an anthropomorphic form with one, three, five, or seven cobra hoods springing from the back of their neck.[5] It is believed that the nāga became a subordinate deity to major Indic religions. However, there are other views as numerous religious structures dedicated to the nāgas are found in different parts of Gaṅgā Valley and other regions that show a preponderance of the nāga cult.[6] The nāgas are worshipped in their reptile and anthropomorphic forms but their realm differs from the human world. In Buddhist literature, they are classed as animals but worthy of worship to soothe their anger and venomous qualities and accentuate their blessings in water and good weather.[7] Aguste Barth summarises the following attributes of nāga worship in India, that is, the worshipping of the most

DOI: 10.4324/9781003601241-4

dangerous reptiles by the human world as deities of water springs and rivers; their mythological origin may be linked to the Ahi of the *Ṛigveda* and a *Vratya* rooted with the qualities of storm, water, and lightning.[8] In the *Vedas*, the nāgas are mentioned as the enemies of the Indra and other Devas as they are aligned with the Dasyus and Asuras. In the *Ṛigveda*, the rivalry of Indra with Ahi, a great snake is well known, and Indra is portrayed as the killer of Ahi.

> *With is a vast destroying thunderbolt, Indra struck the darkling mutilated Vritra; as the trunk of a tree felled by the axe, so lies Ahi prostrate on the earth.*[9]
>
> *The water that delights the minds (of men) flow over him, recumbent on the earth, as a river (burst through) its broken banks; Ahi has been prostrated beneath the feet of the waters, which Vṛtra by his might, had obstructed*[10]

The *Śatapatha Brahmana* informs that Vṛtra was a Nāgaraja.[11] Indra fought with the nāga kings. His conflict with the great Nāga Arbuda is well known in the *Ṛigveda*.[12] The *Mahābhārata* informs that Arbuda was a Nāgarāja.[13] Oldham's argument is acceptable that Vṛtra, Arbuda, Namuchi, Urana, Sushna, Sambara, Pipru, and other chieftains who conflicted with Indra or his followers were tribe chiefs. As Sayana described them in his Vedic commentaries, they were not drought or climatic influencers.[14] These nāgas were elites from the Harappan Civilisation who developed a range of technological developments like making cities, reservoirs, water management systems, etc. They were humans but great worshippers of nāgas. Because of their association with the nāgas, they were represented as the nāgas in the Vedic texts. Their social position declined, but their acumen and knowledge were always a source of inspiration for later traditions. It was believed that they were still ruling the netherworld and many of the great Aryan clans linked their origin with them. The Ikṣvākus are considered to be the Kings of *Patālaloka*.[15] They were the torch-bearers of the Aryan culture in the Gangetic best. Bhagiratha, Rāma, Parśvanātha, Buddha, etc., all were Ikṣvākus, and their contributions to the Indian civilisation are immeasurable. The *Mahāvaṁsa* informs that the Ikṣvākus founded 19 great cities in the different parts of the Indian subcontinent. These cities were Kuśāvatī, Ayojapura, Vārānasī, Kapila, Hatthipura, Ekachakra, Wajiravatī, Madhurā, Aritthapura, Indapattha, Kośāmbī, Konagochi, Roja, Champā, Mithilā, Rājagriha, Takshaśilā, Kuśināra and Tāmralipti.[16] Takshaśilā was founded by Takṣu, son of Bharata and brother of Rāma.[17] The continuous cultural exchanges minimised the mutual hostility between the two groups and the nāgas became well accepted by the Aryans as semi-divines. The *Aitareya Brahmana* informs that Arbuda, an Asura warrior known in the *Ṛigveda* as a son of Kadru and a *riṣi* of a nāga lineage. He was also a composer of Vedic hymns.[18] Balarāma was born as an incarnation of Śesanāga. It is believed that his soul was released from his body after his death as a great nāga.[19] The *Viṣnu Purāṇa* tells that Gargya, father of Indian Astronomy, propitiated Śesanāga to learn astronomical science and knowledge of planets.[20] The *Maitrāyanī Saṁhitā* says that regular homage should be offered to the snakes for the wellbeing of humans.

> *Homage be to the snakes whichever move along the earth, which are in the sky and heaven, homage to be to those snakes which are in the brightness of heaven, which are in the rays of the sun, which have made their abodes in the waters. Homage to those snakes.*[21]

The propitiation of the nāgas under the Buddhist fold was a well-intended mechanism to appropriate the nāga's power of weather control and rainmaking. The Saṅgha became the facilitator of weather control as soon as nāgas became subservient to them. It is well exhibited in the rain-making rituals found in the Mahāyāna texts where the nāgas are represented as the rain's bestower with the Buddha's blessings.[22] In Buddhism, *parittas* were developed as a safety mechanism against snakes. It was based on mutual respect and coexistence. The nāgas are invoked to grant proper rain and material prosperity, and instead of it, the humans are expected to respect the nāgas and pay them homage. In the *Cullavagga* propitiation of four serpent kings (*ahirāja kulāni*), that is, Virūpakkhas, Erāpathas, Chhabyāputtas, and the Kanhāgotamakas have been mentioned (*Virūpakkhehi me mettaṃ, mettaṃ Erāpathehi me, Chhabyāputtehi me mettaṃ, mettaṃ Kaṇhagotāmakehe cha*).[23] The Buddha always took the help of the nāgas during his early wanderings and after becoming the Buddha. Because of that, the Buddhist monasteries are flooded with images of nāgas and their abodes.

Nāgas Association with Water: Buddhist Textual Traditions

In the Buddhist monastic tradition, the sculpture of the nāga is developed generally in the anthropomorphic form with his right hand raised as he is ready to strike. He holds a pot or jar in his left hand, symbolising growth and prosperity.[24] It is a characteristic of Buddhism that some gods and spirits like Brahmā, Sakka, nāgas, yakśas, etc., are considered subservient to the Buddha. Once they came under the suzerainty of the Buddha, they became compassionate and benevolent for the whole world. The nāgas can be the most ferocious and venomous but as soon as they come under the Buddhist fold, they become the bestower of rain and good fortune. With the Buddha's and Dhamma's blessings, they developed all-pervasive kindness and compassion to sustain and protect the world. The legends of Buddha's association with the nāgas were developed as soon as he was born. The *Lalitavistara* informs that on the occasion of Siddhārtha's birth flowers and trees sprouted from the earth and bloomed. Springs and fountains poured perfumed water and aromatic oil. The anthropomorphic nāga damsels appeared in the sky with thousands of articles of worship. Ten thousand nāgas arrived in the sky holding peacock feathers.[25] When Mahāmāyā gave birth to the Siddhārtha in the Lumbinī garden, two Nāgarājas Nanda and Upananda came and showered two streams of cold and hot water for bathing the newborn.[26] Xuanzang says that Aśoka built a stūpa at the place where both the Nāgarājas gave baths to Siddhārtha. To the east of this stūpa, two springs still exist where two nāgas emerged on the birth of Siddhārtha.[27] P. C Mukhopadhyaya has discovered and excavated a large tank adjacent to the temple of Mahāmāyā. This tank is considered the same where the nāgas bathed the prince. Later on, this tank was used for monastic purposes. [28] Many such tanks were also developed at the ancient site of Kapilvastu, modern Tilaurakot, or the Mauja of Tilaura. It is situated in the Siddhārth Nagar district of Uttar Pradesh and extends into Nepal's Lumbinī and Kudan regions. Two tanks are situated in the rampart of Tilaurakot, the larger one near the northeast corner. It contains water throughout the year. The other tank is smaller.[29] When Siddhārtha decided to renounce his home, the Nāgarājas Varuna, Manasvin, Sāgara, Avavatapta, Nanda, and Upananda decided to help him in his quest for *Nibbāna*.[30] Siddhārtha, after prolonged fasting and rigorous asceticism, decided to partake in food, *payas* (milk-rice) was offered by Sujāta. The nāga princess Niranjanā, residing in River Niranjanā, brought for him a jewelled studded throne to sit and partake of the food. Siddhārtha finished milk rice and threw the bowl made of gold in the river. First, it was possessed by the Nāgaraja Sāgara but later on, Śakra took the form of Garuda and took it away. This bowl is still in heaven as an object of worship.[31] The association of the nāgas with Buddhism continued to flourish in the later period. When Siddhārtha started his journey from the bank of River Niranjanā to the Bodhi tree, Nāgarāja Kāla or Kālikā accompanied him. Aśvaghosa says when Nāgarāja Kāla was awakened by the great march of the Bodhisattva Siddhārtha, he paid homage and declared that the Bodhisattva was on the verge of attaining the *Nibbāna*. Being exalted by Nāgarāja, he collected grass from a grass cutter and sat down under the bodhi tree to attain *Nibbāna*.[32] On such a momentous occasion, Nāgarāja Kālikā was accompanied by his queen and many other nāgas. His queen Suvarṅaprabhāsa and other nāga maidens praised the Bodhisattva and offered homage.[33] The *Divyavadāna* says that Aśoka visited the spot where Nāgarāja Kālikā praised the Bodhisattva on his way to the Bodhi tree.[34] Nāgarāja Muchalinda is considered to be the first living being to pay homage to the Buddha. He protected the Buddha for seven days of incessant rain and stormy wind caused by Māra. He used his coils to cover the body and his hood covered his head to protect the Buddha who was still in deep meditation. When the rain was over and the Buddha opened his eyes, Muchalinda loosened his windings from the body of the Buddha and paid reverence to the newly awakened one.[35] The exact number of weeks differs in different literature. The *Nidānakathā* informs that Muchalinda sheltered the Buddha in his sixth week of enlightenment after winning over the temptation of the Mara and his daughters.[36] Lowell Bloss examined these two inherent contradictions to study the linkage of folk tradition with Buddhism. The

Buddha shows two opposite characteristics: he renounced home and wandered in the middle Gangetic region to attain *Nibbāna* and on the other side, after his enlightenment, he returned to the *saṁsāra* and shared compassion with the masses and other living beings, lived among them but always stayed aloof from all kinds of worldly pleasures and material wellbeing. The Muchalinda legend endorses the superiority of the Buddha over the natural forces and spirits, especially nāgas. The legend represents the Buddha as a king and his royal authority exists over the nāgas. But on the other side, nāga as the folk deity with the power to shower rain and storm was acknowledged. His benevolence was necessary for good rain and his placation as he could withdraw from the territory that led to drought and calamities.[37] Nāgarāja Muchalinda was a venomous creature with the immense power to withhold or cause excessive rain. He was absorbed in the Buddhist fold when he shielded the Buddha from the incessant rain.[38] It was an intense attempt by the Buddha to subjugate folk deities like Muchalinda and other nāgas to control and guide the natural forces for the benefit of all sentient beings. This attempt was paramount in the region with a strong influence of the nāga cult. In the stūpas of Amarāvatī and Nāgārjunakoṅda, the Muchalinda legend is vibrantly represented because of the close association of the ruling dynasties like Pallavas, Maitrakas, etc., with the nāgas. It was an exhibition of the Buddha's power for the masses; after him, his relics became directly associated with nāga worship.[39] The sculptures of Amarāvatī show numerous benefactions given by the nāgas to the human world. The legends of Muchalinda, Champeya, and other nāgas are well-exhibited along with the benign presence of the Buddha. Even antagonising relations among shepherds, snake charmers, Garudā, etc., and the nāgas are also depicted, probably as a warning gesture as cruelties against them could invite wrath and disasters.[40] In the Buddhist tradition, Nāgarāja Muchalinda is worshipped in anthropomorphic and aniconic forms. The inscribed image of Muchalinda of the Gupta period (Figure 4.1) from Bodhgayā is a unique example of the human form of Nāgarāja. A stūpa is engraved on the back side of this image. On other occasions, he is shown protecting the Buddha, stūpas, and the Vajrāsana. Many aniconic representations of the Nāgarāja Muchalinda and other nāgas are found across the Buddhist sites of the Indian subcontinent and other parts of Asia.

The image of Nāgarāja Muchalinda (Figure 4.2) protecting the Vajrasana is unique in a carving on a railing pillar of 1st-century CE. It is found in Jagannāth Tekri, Pauni in Maharashtra. When the Buddha began his journey from Bodhgayā to Sārnāth, at first, he was welcomed by Nāgarāja Sudarśana and his followers at Gayā. The Buddha also met another Nāgarāja Kamandalu at Lohitavastu.[41] After the preaching of the *Dharmachakkapavattana Sutta* to the *pañcavaggiyas*, the Buddha returned to Uruvelā and here his first encounter was with the Kassapa brothers (Uruvelā Kassapa, Nadī Kassapa, and Gayā Kassapa). The Buddha visited the abode of Uruvelā Kassapa and sought shelter for a night. The request was granted because the Buddha spent the night where the sacred fire was kept burning and well-guarded by venomous

Figure 4.1 Muchalinda, Bodhgaya, Indian Museum, Courtesy: Sharmishtha Chatterjee.

Figure 4.2 Aniconic Muchalinda protecting Vajrāsana, Pauni, Courtesy: Srikant Ganvir.

nāga. When the Buddha entered the hut the nāga got infuriated and tried to harm him. The Buddha subdued the nāga and after knowing the real knowledge of the Buddha, three Kassapa brothers and their disciples succumbed to join the Saṅgha.[42] In Srāvastī, when the Buddha performed a miracle to subdue the heretics, several dignitaries including King Prasenjita were also present. On this journey, Nāgarājas Nanda and Upananda also accompanied him. They created a lotus throne for the Buddha.[43] The nāga kings and their associates built a bridge over the Gaṅgā to reach Vaiśālī from Rājagriha. Other bridges are built by King Ajātśatru, and by the people of Vaiśālī. The Buddha advised the monks to go by the bridges built by the King and the people of Vaiśālī, but he, along with Ānanda, travelled by the bridge built by the nāgas.[44] The Tibetan legend mentions that two Nāgarājas Girika and Vidyujjvala, a great devotee of the Buddha, lived in Rājagriha. Once, in the presence of King Bimbisāra of Magadha, both paid homage to the Buddha but did not pay attention to the King. Bimbisāra was annoyed by their behaviour and banished both from Magadha. In their absence, there was a shortage of rain in Rājagriha, crops failed, and famine occurred. The King got anxious and asked the Buddha for the reason. The Tathāgata narrated the incident and asked the king to placate them to return to Rājagriha. Bimbisāra propitiated them and also made their shelter at Veluvana.[45] After that, the nāgarājas bestowed kindness to the people of Rājagriha, and the region began to receive a good amount of rain and an abundance of crops.

The legend of Rāmagāma Stūpa suggests how Emperor Aśoka respected and accommodated the request of nāgarāja. Faxian informs when Aśoka visited Rāmagāma Stūpa and intended to excavate the relics of the Buddha, nāgarāja came out and requested him not to do so, as he worships the sacred relics, humans, and animals of the region. A herd of elephants routinely visited stūpa, carrying water from an adjacent pond in their trunk, watering the area, and sweeping it. They also take various types of flowers to pay homage to the relics of the Buddha.[46] Pali and Sanskrit Buddhist traditions accept that the nāgas possessed the body relics of the Buddha at Rāmagāma; however, the legends of both traditions differed from each other. It is understood that Aśoka dug the body relics of the Buddha divided by Droṇa from Rājagriha, Vaiśālī, Kapilvastu, Pāvā, Kuśinagara, Vetthdīpa, and Alakappa.[47] The Stūpa of Rāmagāma was not disturbed by Aśoka. He came to know that the stūpa was underwater in the custody of the nāgas. They took Aśoka to the deep ocean and showed him the relics. The emperor was impressed by the veneration of relics by the nāgas and he decided not to take the relics from them.[48] The *Mahāvaṁsa* gives a different version that the relic stūpa of Rāmagāma was built on the bank of River Gaṅgā. Once in a flood, the relic was sunk and reached the deep ocean where it was discovered and protected by the nāgas. They made the relic stūpa and worshipped it with great veneration. Aśoka did not take it from them as they were devout worshipper of the relics. It happened because the relics from the nāga realm were destined to reach Sri Lanka in the period of King Duṭṭhagāminī.[49] The *Mahāvaṁsa* informs that

Emperor Aśoka requested Nāgarāja Mahākāla to show the human form of the Buddha and his two precursors.[50] Xuanzang also mentions that the nāgarājas like Anavatapta, Muchalinda, and Elapattra came to Brahmana Drona and demanded a share in the Buddha's relics. Drona heeded his request and divided the relics into three parts – one for Devas, one for nāgas, and one for the humans.[51] When Saṅghamitrā was taking a branch of the Bodhi tree to Sri Lanka, the nāgas residing in the mid-ocean tried to snatch but Saṅghamitrā assumed the form of Garuda and subdued them. The nāgas worshipped the Bodhi tree, showered gifts, and consented to guard it.[52] The affable bond developed between Buddhism and nāgas was solely intended to take concern for whole sentient beings. Some of their species are venomous, because of it they are subject to abhorrence and killing. The Buddha tried to remove this incongruity by presenting the cordial reciprocity between nāgas and humans.

The Chinese travellers inform that several abodes of nāgas were considered sacred in different parts of the Indian subcontinent where people thronged on certain occasions to pray for an abundance of rain and good weather. They also paid homage to them to avoid famine, deprivation, and drought.[53] Xuanzang informs that during the period of the Buddha Kaśyapa, a nāga was born in human form, and due to his sacred powers, he was able to control the nāgas of the region to be benevolent and not destructive. Because of his prowess, the nāgas were not able to carry storms or floods, and people were living in great prosperity. They also offered a share of grain to the person for the benevolence done to the people. But soon people forgot to give him grain and in enrage he decided to take rebirth as the nāga and destroy the country that he saved once. After his death, he became Nāgarāja Apalala and caused thunderstorms and deluge to destroy the crops and people. Later on, he was subdued by the Buddha.[54] The *Champeya Jātaka* informs how nāgarāja saved the life of a defeated king who jumped into the river Champa to save his life. He helped the king to restore his fortune and in appreciation, the king erected a jewelled temple on the riverbank. The king used to come to the river bank every year to pay homage to nāgarāja.[55] The treatises on Mahāyāna mention that Nāgārjuna lived in Bhogavatī, a well-known city of the nāgas in the Great Ocean. He visited the palace of the nāgarāja under the sea, a place with abundant gems, jewels, and other material comforts. The nāgarāja had the manuscript of the great Mahāyāna treatise, the *Prajñāpāramitā Sūtra*. It was given to the nāgas by the Buddha to keep it safe till some eminent monk took it. Nāgārjuna was considered the most scholarly monk who could bring this tremendous Buddhist scripture back to the human world.[56] The legends of Rāmagāma and Nāgārjuna suggest that nāgarājas possesses the body relics (Śārīrakāyā) of the Buddha as well as Dharmakāyā, the body of sacred texts like the *Prajnāpāramitā*. The *Harsacarita* repeats the story of Nāgārjuna and a necklace made of precious pearls with more mythical furore. The pearls of the necklace were made out of the tears of the Moon god and were considered the most effective antidote for any deadly venom. Later on, the precious necklace came into the possession of Nāgarāja Vāsuki who gifted it to Nāgārjuna when he visited Nāgaloka to carry the revered manuscript of the *Prajnāpāramitā Sūtra*. He gifted the necklace to the Sātavāhana King and later it came under the possession of King Harsha.[57] The *Ramāyāna* also shows close association with the nāgas, the Earth goddess, and the Ikṣvākus. When Sītā decided to go in the lap of the Earth goddess, she appeared with a golden throne carried by the divine nāgas and took Sītā to the Pātālaloka/Nāgaloka.[58] Xuanzang informs us that hills surround a great lake in the northwest region. It was the abode of nāgas who became enraged at the slightest disturbance. They used to be venomous and led thunderstorms and excessive rain, but monks and lay people, the true followers of the Buddha's Dhamma, could subdue the evil intentions of the nāgas.[59]

Faxian informs that a stūpa was built at the spot near Sārnāth where Nāga Elapattra asked the Buddha some questions and prayed to him to get rid of his birth of nāga.[60] Xuanzang does not mention Sārnāth but informs about a sacred tank of Elapattra in Takshaśilā.[61] Nāga Gopala is considered the guardian of the Buddha's shadow. His abode was on a great mountain in the Kabul Valley. Xuanzang informed that the shadow of the Buddha could be seen here. Nāga Gopāla was a milkman in his previous birth and used to supply milk to a royal palace. Once he forgot to supply the milk and was reprimanded by the King. He became furious and pledged to be a nāga in his next birth to take revenge. After that as nāga he created hail storms and drought and the people of the region became perplexed. When the Buddha was informed, he visited Gopala who submitted to his compassion. On his request, the Buddha left his shadow as his blessings at the cave where Gopāla resided.[62] Xuanzang also informs about Nāgarāja Apalala who resided near a fountain on the bank of River Suvastu in the town of Mangalpura at Udyana. This nāga in the period of the Kasyapa Buddha was a human and his name was Gaṅgi. He was a great charmer and benevolent. He used to teach the snakes not to be venomous, violent, and destructive. Because of his generous nature, the people of the region were happy and prosperous. They used to give tributes and gifts to him. Once they forgot to give tribute to the nāga. He created hail storms and drought. The Buddha accompanied by Vajarapani subdued the power of Apalala and instructed him not to destroy cattle wealth and cattle population. For his survival, the Buddha instructed the people to give some share of produce once in the 12 years.[63]

The Buddha gave due importance to rain and rainwater as they sustain life and give longevity. The *Saṁyutta Nikāya* informs when the Devas asked the Buddha how living beings can survive and live in peace. The Buddha replied that the rainwater is sole energy of life and every living being is sustained by rain.[64] There are five types of clouds: cool clouds (*sītā*), warm clouds (uṇha), storm or thunder clouds (*abha* or *thaneti*), wind clouds (*vata*), and rain clouds (*vassa*).[65] The rainfall has two different natures, that is in the shape of a downpour or large drops (*thulla phusitaka*), generally in the season of monsoon and in the shape of scattered drops (*ekaṁ ekaṁ*), that may be the winter rain.[66] The *Milindapañha* mentions four types of rainfall, the rainy season, the winter season, the two months of Asādha and Sāvana, and the rain out of season.[67] The *Mahāvastu* classifies clouds as *kālamegha* (monsoon clouds) and *akālamegha* (storm clouds).[68] The rain is considered precious for life and linked to the *kuśala-kamma*. The ever-increasing *akuśala kamma* can jeopardise the ecosystem, leading to no rain and drought. The *Vassantarūya Sutta*[69] discusses the five factors that disturb the rain: (i) when the fiery elements develop storms and the frequency of clouds gets scattered, (ii) when winds rage the storm and clouds wither away, (iii) when Rahu gathers water and spills it in the ocean, (iv) when the rain cloud devas became heedless, and (v) the people deviate from *kuśala kamma*. In the early centuries of the common era, the Buddhist monasteries across South Asia developed different types of ceremonies and rituals for rain and rainmaking. In the 4th–5th century, many *sūtras*, *mantras*, and *dhārinis* were developed exclusively dealing with the rain-making rituals, instruction manuals to perform these ceremonies, etc. The core idea of these newly emerged hymnologies was nāga worship, who controls and commands the water and its resources. When people have foreseen drought and famine, the nāgas were placated with suitable offerings as only they could alter the weather and make rain. These rituals and incantations convince the nāgas to help the humans who might have forgotten to worship and give offerings to them. Under the influence of the Buddha's teachings and compassion, they make rain and get rid of drought. The nāgas received incessant offerings for helping the people and in some cases, they were also subjected to subjugation and retribution if they continued to be stubborn.

The nāga legend developed in Mahāyāna and later Buddhism is slightly different from earlier traditions. Here, they became more aqua-oriented and were always described as water spirits residing in the ponds, lakes, rivers, and oceans. They are not depicted as venomous but could invite hail storms, floods, and droughts that could destroy the crops. Their propitiation is for weather control, rain, and to avoid bad fortunes coming through hail storms, floods, and drought.[70] Literature like the *Vajratuṅdasamayakakalparājā*, the *Mahāmegha Sūtra*, etc., assures the creation of favorable weather conditions.[71] The *Megha Sūtra* was developed to invite rain and increase fertility. Many early manuscripts were lost, but a Sanskrit manuscript was discovered in Nepal, and a similar text is also known in China. It is said that the *Megha Sūra* was preached by the Buddha when he was residing in the palace of Nanda and Upananda. The Buddha preached a unique social harmony between humans and animals. The people should not harm the nagas as they give rain and fertility. It narrates:

> *Let the venerable One speak such words of charms that through their utterance here in Jambudvipa, in the latter season and time, in drought, mighty showers may give rain in time of calamity, in time of difficulty, in times of turmoil, in the iron age of lawless people, in time of accident*

> *and misfortune, of disease and death, of the conjunction of adverse planet, let Him work the assuagement of all calamities, accidents and pains. Let the venerable One, of his supreme pitifulness and mercy towards all beings, speak words of spells so formed as to invoke all snakes, destroy all Māras, shield off injuries of all beings and their pains and afflictions and fears, and cause peace and salvation, and mitigate the effects of adverse stars, more over let them stop all the hindrances to rain that have been foretold by the Venerable One and fully creates showers here in Jambudvīpa.*[72]

Vajratuṅda is mentioned as Garuda-linked *dhārini* and denoted as the thunderbolt beak of Garuna.[73] The manuscript is written in Sanskrit in the 5th–6th century CE as a supplement to Varṣapānavidhi (rain-making ritual manual). Five such manuscripts are known in Kāthmandu, Nepal, and two in Tokyo, Japan. The work was also translated into Chinese, Tibetan, and Mongolian languages.[74] The *Sūtra* provides extensive rituals to protect against nāgas and preserve crops. It ensures an abundance of rain and water for good crops and avoidance of drought and famine. The manual informs that recitation of the *maṅtra* invokes the Buddha who would subdue the evil desires of the nāgas. In such invocations, several subsidiary deities, nāgarājas, and sages like Gaurda, Nāgarāja Sāgara, and sage Agastya, etc., were also invited to participate. The texts provide ritual manuals to control the weather to produce abundant crops. It seeks the help of Garuda who is considered the destroyer of the nāgas. Hidas says that it was written after the *Megha Sūtra*, which is straightforward in narration and has fewer complex rituals. The Dharmabhānakas recited it. The *Vajratuṅdasamayakalparājā* was more complicated, ritualistic, and spell-oriented. It adopts many radical means of *maṅtras* and *mudrās*.[75] The *Mahāpratisaramahāvidyārajnī* was written in the 3rd–4th century and informs rituals and *maṅtras* for weather control. In this process, the spell shoal is tied at the top of the flag which calms down all kinds of winds, cold spells, untimely clouds, lightning, and thunderbolts. It eliminates unnecessary instructions and interference, which could demean the crop production. All insects, flies, locusts, and harmful worms can be compelled to leave. All dangerous and malevolent animals, birds, and beasts will vanish from the field. All flowers, fruits, leaves, trees, herbs, crops, etc. will grow well. They will be juicy, tasty, tender, and healthy. The danger of drought will be wiped out, and the rain will abound in season. The great nāgas of the region will ensure rain and productivity.[76] King Harshavardhana's treatise *Nāganaṅda* is based upon the popular theme of Garuda and Nāga rivalry. It also mentions that rain should come at the proper time, as it makes peacocks delighted, dance, and gives an abundance of crops.[77] The *Sarvaduryatiparisodhana Taṅtra* describes a *maṅdala* of the eight Great Nāgas who requested the Buddha to allow them to learn and practice the Buddhist teachings. Even though they were ready to leave their life as nāgarājas, they also vowed to ensure proper rain and good crops. They will also prevent untimely rain. They will destroy all kinds of fears and bestow good fortunes on those who worship the Buddha and Vajradhārā.[78] The legend of Śāntideva is mentioned in Kāthmandu Valley. Once the whole valley was facing severe drought and famine that lasted for seven years. King Gunakara Malla requested Ācārya Śāntideva to perform rituals to placate the nāgas for showering good rain that could save crops and cattle. The king also took a vow that he would regularly visit the place of great Syambhu Stūpa.[79] A Tibetan ritual manual of the 9th–10th century CE discovered at Dunhuang informs two rain-making rituals. In the first, one, Vidyādhara foresees the water *maṅdala* as a sphere which indicates the subjugation of the nāgas and flowing of the water from their bodies. The second rain ritual is more complex. For this ritual, a Vidyadhāra set up a canopy near a pond or a spring. He draws a *maṅdala* with the ochre and also paints the images of the nāgarājas of the four directions on a piece of paper or cloth. He then recites the *Mahāmegha Sūtra* to invite the clouds that come from the mouth of the Nāgaraja.[80] Richard S Cohen says that Buddhism can be understood by the manifestation of local cultures and their absorption in the Buddhist fold. It can be well explained through the broader pattern of mythology, folklore, and the doctrines embedded in the sculptures and the paintings. The dedicatory inscription of Varahdeva found from Ajantā cave no 16 exemplifies such notions. Varahdeva excavated a new dwelling for the nāgarāja near the entrance of the cave on the bank of River Waghera.[81] Early Buddhism relies upon prayers in the form of *parittas* to placate the nāgas. A simple form of *Suttas* recited by the monks on behalf of the whole community was well enough to invite rain and to ward off any mishappenings. In Mahāyāna/Vajrayāna, the nature of worship changed as *maṅtras* and *dhārinis* sometimes invited Garuda to compel the nāgas to come to terms with the deity. The invocation became more ritualistic as well as magic-oriented where incantations and spells became a powerful tool to hallucinate the nāgas for the good fortune of the human world.

The popularity of the image of Agastya is known in the Pāla period in Bengal and Bihar and it became hugely popular in Nepal even after the decline of Buddhism from the northern plain of India. The sculptures and paintings of sage Agastya show his affiliation with Buddhism. Before coming to the north of India, he has a wide presence in the religious paraphernalia of South India during the Chola period and later on. Probably, this cult was carried to North India by the Buddhist missionaries of South India in the early medieval period when Mahāyāna and Vajrayāna became dominant in the Gangetic Valley.[82] In the epics, Agastya is linked to Mitra and Varuna. The legend says that he drank all the ocean water to expand the Aryanisation of South India. In the *Mahābhārata*, he is known as 'Maitravaruni' which indicates his being born from two fathers Mitra and Varuna. Agastya was said to be born from the seed of Mitra and Varuna who were infatuated by Urvaśī. The seed was put in a vessel; because of it, Agastya is known as 'Kumbhayoni'.[83] In Tibetan practice, a monk was appointed hail master to prevent the hail from destroying crops and cattle wealth. It is understood that malevolent spirits caused the hail storm, and it is the duty of the hail master, a lama, to subdue these spirits. Sometimes these spirits were harmed, creating a dilemma for the Buddhist world because it was against the spirit of Mahāyāna.[84] The *Vaṁsavalīs* written in the 18th century says Masyendranāth is a Buddhist deity. Etymologically his name 'Matysa' suggests that he was born from a divine fish and Indra (rain god). The word 'nāth' represents sovereignty. In the Mahāyāna/Vajrayīna traditions, he is known as Karunāmayī Lokeśavara. Because of his name, he is linked to water. Patan and Bungamati celebrate a unique festival in honour of him, and the deity's image is taken on a well-decorated chariot (*ratha-yātrā*). The purpose of this festival is to invite the gods before the monsoon and pray for good rain and crops in the whole Kāthmandu Valley.[85] The Buddhists of Nepal worship Matseyendranāth as a form of Avalokiteśvara rather than a Śaiva deity.[86] The Universal human concern with rain and its disruption is intimately connected with the development of many metaphors of origin, propitiation, and the creation of multiple religious rituals for rainmaking. The source of rain is considered divine and associated with heaven.[87] In the East Asian Buddhist tradition, the dragon has been associated with the symbol of royalty, material well-being, and splendid water sources. In China, multiple rituals were developed to placate the dragons for rainmaking.[88] In Japan, monks have been specially trained to perform the rain-making ceremony with some esoteric practices since the 8th century CE. It was widely believed in Japan that rain-making prayers of the Shingon monks facilitated rain in case of droughts. Because of that, several esoteric rain-making rituals and literature were developed. In these ceremonies, the dragons' or nagas' placation is meaningful. It is believed that dragons possessed power that even Japanese monarchs lacked. They can bestow wealth and increase or decrease rainfall with their divine powers.[89]

Nāgas and Water Management in Indian Buddhist Monastics

Buddhist monasteries were known for their water management. Buddhist monks developed a sustainable approach to water development by harnessing local resources for the monasteries and the people living in the adjoining regions. Water conservation techniques have evolved on existing water sources and the surrounding environment. On the Gangetic plain, water reservoirs were developed along with the monasteries and on certain occasions, some of their water channels connected to the rivers. Cisterns and water tanks were developed on hills in the hill regions for rain harvesting. All these places have nāga sculptures, which were propitiated for good rain and harvest. J. P Vogel noticed that in Mathurā cisterns and water tanks are often adorned with anthropomorphic representations of nāgarāja. His right hand shows a gesture to strike and his left hand holds a sacred pot symbolising life, growth, and prosperity.[90] A Kharoshthī inscription found in the Swat Valley of the 1st century BCE mentions Thera Nora, son of Dati constructed and donated a tank (*puṣakarnī*) to pay homage to all kinds of snakes, on the 20th day of the month of Srāvana of the year 113. Geroge Buhler says he was a monk and probably had a foreign origin.[91] The Chargaon Nāga image inscription of a period of Huviṣka describes the donation of the nāga image and tank (*puṣkarnī*) by two Kuṣāna commanders Senahastin, the son of Piṅdapayya and Bhomuka, the son of Viravriddhi.[92] This nāga image has seven hoods and its coils are visible both in front and back. Here the anthropomorphic representation of nāga is shown by sculpting a human being standing before the polycephalic nāga. The *Kṣudrakavastu* of the *Mūlsarvāstivāda Vinaya* says that the Buddha permits a wealthy donor to build a water tank or cistern that must be adorned with a sculpture of a nāga holding a pot in his hand.[93] A large number of epigraphic records are found in the Gangetic plain and the cave monasteries of western India showing the donation of caves or tanks for monks and nuns.

Uruvelā-Bodhgayā had several natural as well as manmade ponds and tanks. The references of later editions of the *Tipitaka* inform that the name Uruvelā is derived from 'Uru', which means sand, and 'Vela', the waves of sand.[94] It suggests that the village and forest were developed near the river and had frequent floods. Many natural ponds might be created out of it. Xuanzang mentions several tanks and abode of nāgas in the vicinity of Bodhgayā. The Bodhi tree was enclosed by a boundary wall and its main gate opened towards River Niranjanā and the south gate had a large flower tank. Outside the Bodhi tree's southern entrance, a large tank measuring 700 paces was situated. It was home to many nāgas and fish. A Brahmana built this tank. In the vicinity of this tank was another tank created by the Sakka where the Buddha washed his robes. To the east of the Sakka's tank, the pond of Nāgarāja Muchalinda was situated. Here, the Buddha sat for seven days and the Nāgarāja protected the Buddha from incessant rain by coiling him with his body and hoods. On the opposite of the bank, Nāgarāja Muchalinda resided.[95] In the south of the Muchalinda's tank is where Kasyapa Buddha went to protect the Buddha from the flood and drowning. It is presumed that the Kassapa brothers of Uruvela created thunderstorms with their occult power, but the Buddha subdued the rain, thunder, and flood. When Kasyapa reached by boat to rescue the Buddha, he found him walking on dry sand, and a flood on all sides. He respected his power and returned. Similarly, outside the eastern gate of the Bodhi tree, there is the abode of blind Nāgarāja; Siddhārtha on his way to Prāgabodhi met him and his eyesight returned. He predicted that the Bodhisattva was going to be the Buddha soon.[96] Alexander Cunningham identifies the Buddhokhar Tāl as the tank dug by the Brahmana. It is situated 300 feet to the south of the courtyard of the main temple and measures 400 square feet or 1,629 square feet in circuit. Further south, another tank may be the Sakka's tank built for the Bodhisattva to bathe and wash his robes. It is now known as Mahap Tāl, measuring 300 feet long and 250 feet wide. To the south-west of the Sakka's tank, lies the tank of the Nāgarāja Muchalinda. It is now known as Teskar or Taksal Tāl. It is 600 feet long and 500 feet wide. In the vicinity, another great tank belongs to the village Urel, a representative of ancient Uruvelā.[97] These tanks may have been natural when the Buddha attained *Nibbāna*. With the growth of monastic settlements around the Bodhi tree, most of these tanks were renovated and used for monastic requirements. The archaeological remains of Bakhra are situated approximately three kilometres northwest of Vaiśālī. Here, adjacent to a pillar of Aśoka, a tank measuring 200 feet from east to west and 150 feet from north to south is situated. Alexander Cunningham identifies it with Markata-hṛda or the monkey tank where the monkeys offered honey to the Buddha.[98] It was a tank made explicitly for bathing for monks and nuns. A special arrangement is made such that bathing ghats on the other side are not visible to the person bathing at one end. This unique mechanism was developed to create a separate space for monks and nuns residing in the separate monasteries at this place.

Buddhism took its cultural milieus wherever it went in the early expansion phase. The early monastic settlements in Sanchi and western India show that Buddhist monks adopted nāga worshipping and integrated it with local water and rain-making cults. Buddhism made subtle inroads in the Avanti region during the Mauryan period. The monks developed their monasteries on hills and plains and introduced new crops and water management techniques. Significantly, water management was always supported by the nāga cult. The anthropomorphic image of the nāgarājas has been found in different places in Sanchi, Andher, and other areas from the early centuries of the common era. The image displayed in Sanchi Museum belongs to the 4th–5th century (Figure 4.3).[99]

Julia Shaw examined the approximately 35 sites of the Sanchi Buddhist Complex covering an area of 750 square kilometres. Most of these sites are in Satdhara, Sonari, Morel Khurd, and Andher, where free-standing nāga sculptures are found from the 1st century BCE. The nāga cult and the dams for irrigation and monastic use began to develop in the 3rd century BCE. In Aśoka, many dams were raised to facilitate paddy cultivation and lasted until the 12th century CE. The cultural topography of the region suggests that the Buddhist Saṅgha acclimatised themselves to the region's social, economic, and religious spectacle, which led to the development of a patronage network of pan-Indian as well as a pan-Asian character. Buddhist establishments and the raising of water dams in the Sanchi region show how religion was responsible for changing the economic ecology of the region. Raising the nāga sculpture along with the water dams or cisterns could influence the monsoon rain, an essential water and land management source. It was a pragmatic mechanism to alleviate the suffering (*dukkha*).[100] John Marshall suggests that the earliest date of raising water dams in the hill goes back to the 3rd century BCE but started regularly since the 2nd century BCE.[101] The structure of dams in the Sanchi region is quite similar in shape and size. The dams were built of earthen material with the capping of stone on the upstream side. The height of dams varies from one meter to six metres and the length varies from

Figure 4.3 Nāgāraja, Sanchi Museum, Image by author.

80 to 1,400 metres. They were raised at a considerable height to regulate the flow of reservoir volume with the volume of runoff from the upstream catchment of each side.[102] These reservoirs have water volumes ranging between 0.03 and 4.7 $m^3 \times 10^6$, that is 30,000 to 47,00,000 m^3. The dams which are built on gently sloping terrain were used for upstream irrigation and dams raised on deeper valleys were built to irrigate downstream irrigation. The downstream irrigation is embedded with flood-control spillways and sluices.[103] Nāga sculptures linked with rain and fertility are found throughout the Sanchi area. Their spatial and temporal distribution in this area show important characteristics of religious change, especially about the dynamics between Buddhism and local agricultural threads. The establishments of nāga shrines confirm this model and endorse the hypotheses present elsewhere.[104] It is also believed that Buddhism amalgamated the local deities like the nāgas and yakṣas in their fold which helped them to proselytise the local population into the Buddhist fold. The Buddha's accommodation, appropriation, and interface with the nāgas, yakṣas, and their folk traditions can be visualised from Buddhist texts, sculptures, bas-reliefs, etc. It is also possible that their non-religious work for public causes like water management, improved agricultural networks, and medical facilities led to encouraging the people to embrace Buddhism. The nāga was introduced into the Buddhist ritual practices as a means of legitimising the Saṅgha's presence in the area of

its expansion. The purpose of the Buddha's subjugation of the nāgas was to exhibit the different notions the nāga folk cult carried to the masses. The nāgas can be venomous and interrupt the monsoon rain and the maker of the deluge. When they became subordinate and protectors of the Buddha or the Buddha's relics, their munificence and benevolence as the guardians of subterranean and sub-aquatic treasures manifested under the Dhamma's influence. Because of that, they were considered the protectors of the stūpas and vihāras, as well as the large populace depending upon them at places like Sanchi, Pitalkhora, etc.[105] R. S. Cohen summarises that the appropriation of the nāga cult in Buddhism was due to their authority over a particular region and its resources. Their subordination in the fold of Buddhism helped the Saṅgha to localise and mingle with the local population. The representation of nāga in cave number 16 at Ajantā was not because he just embraced Buddhism or became the protector of the Buddha but as a sole beneficiary of the Buddha's protection. He was a natural inhabitant of that place and under the tutelage of the Buddha, he became more protuberant and strengthened.[106] The cause of the legitimisation of nāga's presence with the Buddha at this cave was the water sources. In the hills, where seasonal rains are the only source of water, the blessings of nāga became important for the local population as well as the Saṅgha. Ony dry spell of monsoon will jeopardise the local socio-economic setup that the Saṅgha and lay society could not sustain.

Similar experiments were done in the hill monasteries of western India. The caves of the Western Ghats faced long spells of dry season as well as an abundance of rain in the rainy season. At one end, they faced scarcity of water in the summers and at the other end green landscape with a lot of rain and water springs in the rainy season. The monasteries of Ajanta, Kanehri, Bhaja, Pitalkhora, etc., faced similar conditions and developed a unique kind of nāga worship linked to their water management. They conserved the excess water of the rainy season through rain harvesting. For that, they developed a highly efficient and mechanised water management and hydraulic system and people were encouraged to donate a cistern or tank for the monasteries. Kanheri or Krisnagiri lies in the vicinity of Mumbai and its Buddhist cave complex comprises more than a hundred caves. The beginning of the cave monasteries may be from the 1st century BCE onwards and lasted up to the medieval period. The site has a unique characteristic with its water management system, agricultural land, and satellite settlements. The cave settlements are spread over the Borivali National Park and most of the caves are situated on the southern hill with well-connected water cisterns. The northern hill has fewer caves. Water streams flow between the northern and southern hills and water from all hills merge at the eastern point. The monks developed the catchment area of streams and small reservoirs by raising two walls between the northern and southern hills. On the top of the southern hills, there were five water tanks. Most cisterns and tanks are connected and with a small network of sluices. The rainwater is harvested in these tanks and cisterns for drinking and other purposes. On the eastern side of the hill

Gomukha, water from a natural spring was collected in a small tank.[107] The names and donations of cisterns and tanks are known as *podhī*, *pānīyapodhī*, *pānīyaka deyadhamma*, *pānīyaka bhājanam*, *nhāna podhī*, and *talāga*. 21 inscriptions are found in the different caves referring to donations of cisterns and tanks at Kanherī. The inscription found at cave no 5 informs that the wife of Vaśiṣthiputra Pulamāvi and daughter of Rudradāmana also donated a cistern at Kanheri.[108] Two types of water cisterns, that is, in an open courtyard and the recesses along the walls of the courtyard, are carved out. In some of the caves, cave no 2, two tanks were excavated adjacent to each other. The larger tank was for collecting rainwater and water from the natural springs and was equipped with a wooden lid to cover it. It might serve the purpose of drinking water. The other tank is more prominent and rectangular and must have been used for sanitation, washing, and baths. The water management of the hills is embedded with a network of small sluices one inch to eight inches deep and two inches to 10 inches wide. It started from the smaller tanks on the southern hill and connected the cisterns of the uppermost caves. When water is filled in these cisterns, the water inlets connected to these cisterns collect all water from the northern and western slopes of the southern hill and channelise it to the next level of caves. When these cisterns get filled, the water was further diverted to lower levels of caves.[109] The reciprocity and understanding between the monks and nāgas are pretty visible here with their number of sculptures. The most aesthetic sculpture of nāga is on the entrance of cave no 3. Here the image was directly placed above one of the water cisterns.[110] The mechanism developed by the monks and nuns is still the source of water conservation for the local villagers residing in the nearby forest range. An expanded network of dams and water channels is also found at Pitalkhora in the Sahayadri–Shatamala range on the northern fringe of Aurangabad. Barrage, desilting pits, and water channels are essential features of this ancient hydraulic system.[111] The water management at Pitalkhora is considered the oldest in western India, around the 2nd–3rd century BCE. A multi-headed nāga with a water spout in each mouth is carved outside cave no 3, near the cistern that collects water from the hill (Figure 4.4).

In the rainy season, water comes through this narrow channel and passes through the spouts carved into the nāga's multiple mouths. The monks of this region acclimatised to the local environment and properly understood water scarcity. They developed an efficient water management technique that transformed the local socio-cultural fabric.

Sankissa is one of the eight great places (*atthamahātthāni*) in Buddhism and here, the Buddha descended from the Tavatimsa heaven along with the Devas Brahmā and Sakka, after preaching to his mother Mahāmāyā. The remains of many stūpas and vihāras are discovered in this place. Faxian mentions the prevalence of nāga worship in the monasteries of Sankissa; his abode was near the main tank of the stūpa complex. He was popularly known as 'White Ear' Nāgarāja. The monastic communities and the lay people thronged to worship him, especially in the month of the *Sāvana* for good

Figure 4.4 Multi-headed nāga with spout, Pitalkhora, Courtesy: Srikant Ganvir.

rain and avoidance of drought, plagues, and calamities. The monks of the monasteries have raised a chapel for him.[112] Xuanzang informs that the nāgarājas of the water tank protect the Buddhist edifices like stūpas, vihāras, and stairs symbolising the move towards heaven, and a stone pillar raised by Aśoka to celebrate the occasion.[113] The 'White Ear' Nāga tank is identified in the southeast of the remains of the monastic structure. This naga is known as Kārewar, and the tank is called Kāndaiya Tāl. Milk is offered to the nāga every day of the month of Vaisakha to celebrate the festival of Nāga Panchamī. Cunningham argues that the nāga worship in the monastery of Sankissa is not associated with the Nāga Panchamī prevalent in most parts of North India but is specifically dedicated to the prayer for the abundance of rain and avoidance of drought. In the month of Vaiśākha, when water scarcity looms in the Northern plain, the people come here and offer milk for rain.[114] Initially, the tank may be natural but with the development of monasteries, it was enlarged and adequately maintained. Along with the worship of the nāga cult for rain, the water of the tank also served the purpose of cleaning, washing, and sanitation. B. R Mani found evidence of habitation since the PGW period, and the place was well settled during the period of the Buddha, as numerous pieces of evidence of sherds of NBPW have been discovered in the monastic region as well as in the neighbouring settlements. A temple of Bisharī Devī covers the main stūpa.[115] Bisharī Devī was a parochialised form of worshipping of the Nāga of Sankissa. After the decline of Buddhism in this part of India, nāga worship continued but in different forms. The deity

Bisharī became the guardian of the nāga of the tank and her worshipping along with the nāga is now not only for the rain but also for protection from venomous snakes.

Shravasti is another great Buddhist site in the middle Gaṅgā Plains where water management and cisterns were well developed. Despite being well connected with River Rapti, Shravasti had developed a form of water tanks in the town and the monasteries.[116] Japanese archaeologists led the excavation of Shravasti for five seasons from 1986 to 1996 and found that the ancient site had a circumference of 5.23 kilometres. It was situated on the old bed of River Rapti. The city shows habitation since the early historic period and the rampart was made in the period of the Buddha, that is, the NBPW age.[117] During the excavation of the monastic site Sahet (ancient Jetavana), a large tank and monastic structures were excavated. The tank was square in shape, measuring 27.30 metres from east to west and 24.70 metres from north to south in its early phase, and was probably used for bathing purposes by the monks and nuns of the monastery. It was built with the monastery's foundation but was subject to renovation and repair in different periods due to the regular occurrence of floods. The brick in the early stratification measures 48.0×24.0×8.0 metres. The wall of the tank is slightly tilted inward to avoid the pressure caused by soil or the occurrence of floods, a regular feature of this region. The tank was used for a long time and later on, one inlet on the southeast corner and one outlet channel on the west were built to draw, drain, and clean the water. The ramps were also developed on all four sides of the enclosing walls. Wooden logs were used to make the ramps and give them support.[118] The Jetavana monastic site continued till the 12th century CE. The donative record dated 1176 of the Vikrama eras or 1118 CE informs that Vidyādhara built a vihāra for the monks in Jetavana. His father Janaka was a minister of King Gopāla of Gādhipura or Kanyakubja and served as King Madanapāla's minister. The inscription calls Gopāla the lord of Gādhipura but this title was not given to King Madanapāla. It shows that this area was under the Pālas of Bengal but later on, annexed by the Gahadavālas.[119] The patronage of Jetavana continued when the Gahadavālas conquered Kannauj and the neighbouring regions. The copperplate inscription of the 12th century CE of King Govindrachandra of Kannauj informs the grant of six villages to Jetavana Vihāra. The victorious king Govindachandra, after taking the sovereignty of Kanyakubja, issued grants to the Jetavana monastery. Under the guided influence of Saugāta Parivrajaka Mahāpaṇdita Śakyaraksita of Utkala and his disciple Saugāta Parivrajaka Mahāpaṇdita Vagisvararaksita of Choda country, he granted the revenue of six villages Vada-chaturasiti, Pattana, Upalaunda, Vavvahali, Ghoshadi of Meyi, and Payasi attached to Potthivara on Monday, Purnima *tithī* of the month of Asadha in the Purvasaddha nakṣatra in the year of Samvat 1186 with rights of water and dry land, mines of iron and salt, ponds of fish, garden, bushes, grass, and pasture land to the Buddhist monks (*Śākyabhikkhus*) of whom Buddhabhattarika is the chief residing in the great vihāra of

Jetavana.[120] The evidence of thick layers of charcoal belonging to the early medieval period suggests that the monastery was destroyed by fire.[121] Its destruction was by the Islamic invasions as the area was under continuous war and plunder by the Turks. Though the water management, as well as the monasteries, were destroyed, nāga worshipping never declined in this region and became a continuous thread of Hindu tradition where snakes were worshipped to get rid of their venomous bites.

Nālandā Mahāvihāra was a great seat of Buddhist knowledge that reigned the academic world of Asia for thousands of years. I-Ching says that the name was derived from a nāga king, Nāga Nanda.[122] It is difficult to accept that Nālandā was named after the nāga king Nanda but the connotation of the land of Magadha with the serpent cult is very old. The presence of many images of nāgas and nāgīs in the archaeological museum and surrounding areas indicates that the nāga cult was popular in Nālandā. An image of nāgarāja carved on a granite stone was discovered in this complex and was preserved in the museum located at this site. The seven-hooded nāgarāja in an anthropomorphic form is seated in *padamasana-mudrā* on a lotus throne. He is adorned with rich ornaments like two layers of necklace, armlets, *kuṅdalas* in ears, a girdle, and a crown.[123] The monasteries of Magadha began to develop water-harvesting mechanisms since their foundation, and one of the early experiments was done on the hilltop monastery at Giriyaka. An oblong water reservoir has been found at the receding end of the higher slope of Giriyaka hill in Rājgir, within 50 feet of the main stūpa. This tank was excavated from two sides of the natural rock of the hill and plucked up with stone blocks on the other two sides. It was one of the earliest experiments in the land of Magadha. It was not easy for the monasteries on the hills to get water, so such monastic establishments developed mechanisms of making cisterns and reservoirs for rain harvesting. The masonry and use of stone blocks suggest that its excavation was pre-Mauryan with the support of the state patronage. It provides empirical information for re-assessing the development of monastic settlements, water management, and the state's role. The reservoir offers interesting indications of how kings played an active role in setting up reservoirs for the survival of the monasteries.[124] The monks of Nālandā developed extensive reservoirs, water channels, and wells. The success of such a great endeavour was based upon the sustenance and preservation of water. Great water engineering directed substantial amounts of water necessary for monastic use, bathing ponds, and water reservoirs, which helped preserve water to avoid scarcity. The monks utilised their knowledge of climate science, mathematics, and geometry to deal with the challenges of constructing water channels, preserving natural ponds, and constructing new ponds and dikes (*ahara-payne*) that controlled the flow of water. Magadha's weather conditions were highly unpredictable. While sometimes it would rain heavily, drought conditions would prevail on other occasions. In such a scenario, the monastic establishment would usually be faced with a potential famine or flood situation, which was a catastrophe as it depended on agriculture. Despite such a dwindling situation, Buddhism flourished in Nālandā and the adjoining regions. This was primarily because the monks and the Saṅgha tried to find solutions for all the problems and in the process helped the Buddhist monastic order not only to survive but also to prosper. It seems that the palaeo-channel connecting River Panchana to the Dīghī Pokhar was planned and executed by the Mauryas during this long period of famine. The local monks invented the' ahara-payne' system in Nālandā and other parts of Magadha. It may have been invented to help the agrarian system to function smoothly. After all, the monastic order's survival depended on a stable agrarian foundation.[125] Water management in the monastic architecture of Nālandā includes standing structures like ponds, wells, and drains and underground features like pipes, drains, and sewers. The remains of the Nālandā water system are perhaps the best-known examples that describe the use of sources of water by the Buddhist monastery. The study of these structures requires an understanding of the technology used in their construction and the operation of the water system. The topography, scale, and complexity of the water management system at Nālandā required an exceptional level of technology. I-Ching has discussed this issue at length. The storage, preservation, and cleanliness of water were important and routine works for the monks. According to him, water and pools were in abundance. The monastic complex of Nālandā had ten great pools. Although these pools were rain-fed, the level of cleanliness and hygiene was very high. On the four sides of the bathing ponds, *sala* trees were planted (which today have grown to almost 40 to 50 feet), probably to provide relief from heat. The monks, every morning, were informed of a bathing hour through the sounding of the bell. Sometimes 100 and on occasion almost 1,000 monks went together to take a bath with their bathing sheets.[126] Despite such vast paraphernalia of water management, the Saṅgha always believed that the blessings of the nāgas are always important for good rain and crops and they kept propitiating them till the existence of the Mahāvihāra.

Notes

1 T. V. Mahalingam, *The Nāgas in Indian History and Culture, Sir Subramanyam Ayyar Endowment Lecture, University of Madras*, Madras: G S Press, 1965, p. 7.
2 J. P. Vogel, *Indian Serpent-Lore or The Nagas in Hindu Legend and Art*, Delhi: Gyan Publishing House, 2021, pp. 30–31 (reprint).
3 Mircea Eliade, *Patterns in Comparative Religion*, London: Sheed and Ward, 1958, p. 211.
4 B. N. Puri, 'Irrigation and Agricultural Economy in Ancient India', *Annals of the Bhandarkar Oriental Research Institute*, vol. 48 no. 9, 1968, pp. 383–390.
5 Lowell W. Bloss, 'The Buddha and the Naga: A Study in Buddhist Folk Religiosity', *History of Religions*, vol. 13, no. 1, 1973, p. 37.
6 H. Hartel, *Excavation at Sonkh: 2500 Years of a Town in Mathura District*, Berlin: Dietrich Reimer Verlag, 1993, pp. 10–16; G. C. Chandra, 'Excavation at Rajgir', *ASIAR*, 1935–1936, Delhi: ASI, 1938, pp. 52–54.
7 Vogel, *Indian Serpent-Lore*, p. 3.

8 Auguste Barth, *The Religions of India*, Boston: Houghton Mifflin and Company, 1882, pp. 266ff.
9 *Ṛgveda*, I.32.5.
10 *Ṛgveda*, I.32.8.
11 *Śatapatha Brahmana*, I.6.3.9.
12 *Ṛgveda*, I.51.6.
13 *Mahābhārata, Sabhāparva*, p. xxi.
14 C.F Oldham, *The Sun and the Serpent, A Contribution to the History of Serpent Worship*, New Delhi: Gyan Publishing House, 2020, pp. 42–43, reprint.
15 *Viṣnu Purāṅa*, IV.4.378.
16 *Mahāvaṁasa*, p. XXXV.
17 *Viṣnu Purana*, IV.4.219.
18 *Aitareya Brahmana*, VI.1.1.
19 *Viṣnu Purāna*, V.36.11.
20 *Viṣnu Purāna*, II.5.206.
21 *Maitrāyanī Saṁhitā*, II.7.15.
22 C. Bendall, *The Megha Sūtra, Journal of Royal Asiatic Society*, 1880, pp. 237–240.
23 *Cullavagga*, V.6.
24 J. P. Vogel, 'Nāga Worship in Ancient Mathura', *ASI Annual Report*, 1908–1909, pp. 159–163.
25 Rajendralala Mitra, *The Lalitavistara or Memoirs of the Early Life of the Śākya Sinha*, Calcutta: Asiatic Society, 1882, pp. 110–111.
26 *Lalitavistara*, p. 124.
27 Tomas Watters, *On Yuan Chwang's Travels in India (AD 629–645)*, 2 vols., New Delhi: Low Price Publication, 2004, vol. II, p. 14, reprint.
28 P. C. Mukhopadhyaya, *A Report on a Tour of Exploration of the Antiquities in the Tarai, Nepal, the Region of Kapilvastu, during February and March 1899*, Calcutta: Office of the Superintendent of Government Printing, India, p. 38.
29 Debala Mitra, *Excavations at Tilaura-Kot and Kondon and Explorations in the Nepalese Tarai*, Calcutta: Archaeological Survey of India, 1972, pp. 10–11.
30 *Lalitavistara*, I, p.204.
31 Samuel Beal, *The Romantic Legend of Śākya Buddha, A Translation of Chinese Version of the Abhiniṣkramanasūtra*, New Delhi: Samyak Prakashan, 2011, p. 195, reprint.
32 *Buddhacarita*, XII.113–116.
33 *Jātāka*, no. 70, 72; *Mahāvaṁsa*, xxxi.83.
34 *Divyāvadāna*, pp. 392 ff.
35 *Mahāvagga*, I.3.3.
36 T. W. Rhys Davids. *Buddhist Birth Stories*, London: Trubner, 1880, p. 109.
37 Bloss, 'The Buddha and the Naga: A Study in Buddhist Folk Religiosity', pp. 36–37, 50.
38 Julia Shaw, 'Archaeology of Buddhist Propagation in Ancient India: "ritual" and "practical" Models of Religious Change', *World Archaeology*, vol. 45, no. 1, pp. 83–108, p. 96, Doi:10108/00438243.2013.778132.
39 Bloss, 'The Buddha and the Naga: A Study in Buddhist Folk Religiosity', pp. 50–51.
40 C. Sivaramamurti, *Amaravati Sculpture in the Madras Government Museum, Bulletin of the Madras Government Museum*, Madras: Thomson and Company Limited, 1956, p. 73.
41 *Mahāvastu*, vol. III, pp. 324–327; *Buddhacarita* XV.96.
42 *Mahāvagga*, I,15.
43 W. W. Rockhill, *The Life of the Buddha and the Early History of His Order*, London: Trubner and Company, p. 79.
44 *Divyavadāna*, pp. 55ff.

45 Vogel, *Indian Serpent-Lore*, p. 118.
46 Samuel Beal, *Travels of Fah-Hian and Sung-Yun from China to India (400 A.D. and 518 A.D.)*, Delhi: Low Price Publications, 2005, pp. 90, 91 (reprint).
47 John S. Strong, *Relics of the Buddha*, Delhi: Motilal Banarasidas, 2007, pp. 126–127.
48 Strong, *Relics of the Buddha*, p. 219.
49 *Mahāvaṁsa*, p. 211.
50 *Mahāvaṁsa*, V. 87–94.
51 Watters, *On Yuan Chwang's Travels in India*, vol. II, p. 42.
52 Watters, *On Yuan Chwang's Travels in India*, XIV.17–23.
53 Beal, *Travels of Fah-hian and Sung-yun from China to India*, p. 67; Samuel Beal, *Life of Hieun-Tsiang by the Shaman Hwui Li*, London: K. Paul, Trench & Trubner, 1911, p. 137.
54 Watters, *On Yuan Chwang's Travels in India* II, pp. 320–321.
55 *Jataka*, no. 506.
56 Jan Westerhoff, *The Golden Age of Indian Buddhist Philosophy*, Oxford: Oxford University Press, 2018, p. 91.
57 *Harsacarita*, pp. 250ff.
58 *Rāmāyana, Uttara Kanda*, XCVII, 17–20.
59 Thomas, *On Yuan Chwang's Travels in India*, vol. I, pp. 66–67.
60 James Legge, *A Records of the Buddhistic Kingdoms Being an Account by The Chinese Monk Fa-Hien of His Travels in India and Ceylon (AD 399–414) in Search of the Buddhist Books of Discipline*, Oxford: Clarendon Press, 1886, p. 96.
61 Watters, *On Yuan Chwang's Travels in India*, vol. I, pp. 199, 241.
62 Watters, *On Yuan Chwang's Travels in India*, vol. II, pp. 184, 193.
63 Watters, *On Yuan Chwang's Travels in India*, vol. I, p. 228.
64 *Saṁyutta Nikāya*, I. 81.
65 *Saṁyutta Nikāya*, III. 254.
66 *Aṅguttara Nikāya*, I. 243.
67 *Milindapaṅha*, IV. 1.36.
68 *Mahāvastu*, II. 34.
69 *Aṅguttara Nikāya*, III. 243.
70 Vogel, *Indian Serpent-Lore or The Nagas in Hindu Legend and Art*, p. 93.
71 Robert Decaroli, 'Snakes and Gutters Naga Imagery, Water Management, and Buddhist Rainmaking Rituals in early South Asia', *Archives of Asian Art*, vol. 69, no. 1, 2019, pp. 1–2.
72 Cecil Bendall, 'The Megha Sutra', *The Journal of the Royal Society of Great Britain and Ireland*, New Series, vol. 12, no. 2, 1880, p. 299, www.jstore.org/stable/25196850/.
73 Laurence Waddel, 'The Dharani Cult in Buddhism, Its Origin, deified Literature and Images', *Ostasiatische Zeitschrift*, vol. I, 1912, p. 175.
74 Gergely Hidas, *A Buddhist Ritual Manual on Agriculture Vajratuṇda-samayakalparāja*, Critical Edition and Translation, Berlin/Boston: Walter de Gruyter, 2019, pp. 4–6.
75 Hidas, *A Buddhist Ritual Manual*, pp. 7–11.
76 Gergely Hidas, *Mahāpratisaramahavidyārajnī, The Great Amulet, Great Queen of Spells, Introduction, Critical Edition, and Annotated Translation*, New Delhi: International Academy of Indian Culture and Aditya Prakashan, 2012, p. 224.
77 Andrew Skelton, *How Nagas were Pleased by Harsha and the Shattered Thigh by Bhasa*, New York: New York University Press, 2009, pp. 222–223.
78 Tadeusz Skorupski, *Sarvadurgatiparishodhanatantra, Elimination of All Evil Destinies*, Sanskrit and Tibetan Text with an Introduction, English Translation and Notes, New Delhi: Motilal Banarasidas, 1983, pp. 58, 206.

79 Max Deeg, *Miscellanae Nepalicae Early Chinese Reports on Nepal, the Foundation Legend of Nepal in its Trans-Himalayan Context*, Lumbini: Lumbini International Research Institute, 2016, pp. 195–206.
80 Hidas, *A Buddhist Ritual Manual* pp. 15–16.
81 Richard S. Cohen, 'Nagas Yaksini, Buddha: Local Deities and Local Buddhism at Ajanta', *History of Religions*, 1996, vol. 37, no. 4, 1998, pp. 373–375.
82 Carol Radcliffe Boton, 'Images of Agastya in Nepal', *Artibus Asiae*, 1991, vol. 51, no. 1/2, p. 75.
83 *Matsya Purāṇa*, 61.20–32; *Mahābhārata*, 3.98, 7.157.
84 Anne C. Klein and Khetsun Sangpo, 'Hail Protection', in *Religions of Tibet*, ed. Donal S. Lopez Jr., Princeton: Princeton University Press, 1997, pp. 538–547.
85 Prem K. Khatry, 'Rain for the Drought: An Anthropological Inquiry into the Structure of the Buddhist Festival in Kathmandu Valley', *Contribution to Nepalese Studies*, vol. 23, no. 1, 1996, pp. 89–90.
86 S. J. Locke and K. John, *Karunamaya: The Cult of Avalokieśvara Masyendranāth in the Valley of Nepal*, Kathmandu: Sahyogi Prakashan, 1980, p. 405.
87 Brian O Ruppert, 'Buddhist Rainmaking in Early Japan: The Dragon King and the Ritual Careers of Esoteric Monks', *History of Religions*, vol. 42, no. 2, 2002, p. 143.
88 M.W de Viser, *The Dragon in China and Japan*, Amsterdam: Johannes Muller, 1913, pp. 7–8.
89 Ruppert, 'Buddhist Rainmaking in Early Japan', pp. 145–147.
90 J. P. Vogel, 'Naga Worship in Ancient Mathura', *ASI Annual Report*, 1908–1909, pp. 159–163.
91 George Buhler, 'A New Kharoshthi Inscription from Swat', *The Indian Antiquary*, vol. XXV, 1896, pp. 141–142.
92 Ludar's List no. 1419.
93 Robert Decaroli, *Haunting the Buddha*, New York: Oxford University Press, 2004, pp. 83–840.
94 R. Spence Hardy, *A Manual of Buddhism, in its Modern Development*, London: Partridge and Oakey, 1853, p. 164.
95 Watters, *On Yuan Chwang's Travels in India*, vol. II, pp. 113, 127–128.
96 Watters, *On Yuan Chwang's Travels in India*, vol. II, pp. 132–134.
97 Alexander Cunningham, *Report of Tours in the Gangetic Provinces From Badaon to Bihar in 1875–76 and 1877–78*, New Delhi: Archaeological Survey of India, 2000, pp. 146–147, reprint.
98 Alexander Cunningham, *Four Reports Made During the Years 1862–63–64–65*, New Delhi: Archaeological Survey of India, 2000, vol. I, pp. 62–63.
99 Sanchi Museum, Acc. No. 2858.
100 Julia Shaw, 'Archaeologies of Buddhist Propagation in Ancient India: "Ritual" and "Practical models" of Religious Change', *World Archaeology*, vol. 45, 2013, pp. 84–86, DOI:10.1080/00438243.2013.778132.
101 John Marshall, *The Monuments of Sanchi*, Calcutta: Survey of India, 1940, p. 13.
102 Julia Shaw, John Sutcliffe, Lindsay Loyal Smith, Jean-Luik Schwennenger and M. S. Chauhan, 'Ancient Irrigation and Buddhist History in Central India: Optically Stimulated Luminescence Dates and Pollen Sequences from the Sanchi Dams', *Asian Perspectives*, vol. 46, no. 1, 2007, pp. 166–172.
103 Julia Shaw and John Sutcliffe, 'Ancient Dams and Buddhist Landscapes in the Sanchi Area: New Evidence for Irrigation, Land Use, and Monasticism in Central India', *South Asian Studies*, vol. 21, 2005, pp. 1–24.
104 Julia Shaw, 'Naga Sculptures in Sanchi's Archaeological Landscape: Buddhism, Vaishnavism and Local Agricultural Cults in Central India, First Century BCE to Fifth Century CE', *Artibus Asiae*, vol. 64, no. 1, pp. 7–10.

105 Shaw, 'Archaeologies of Buddhist Propagation in Ancient India: "Ritual" and "Practical models" of Religious Change', pp. 95–96.
106 R. S. Cohen, 'Naga, Yaksini, Buddha: Local Deities and Local Buddhism at Ajanta', *History of Religion*, vol. 37, pp. 377–378.
107 Suraj A. Pandit, 'Water Management System at Kanheri', *Journal of Asiatic Society of Mumbai*, 2010, pp. 28–29.
108 S. Gokhale, *Kanheri Inscriptions*, Pune: Deccan College, 1991, pp. 47. 62, 91.
109 Pandit, 'Water Management System at Kanheri', pp. 30–32.
110 Decaroli, *Haunting the Buddha*, p. 9.
111 S. K. Mitra, 'Ancient Barrage at Pitalkhora', in *Puraratna-emerging Trends in Archaeology, Art, and Anthropology, Conservation, and History, in Honour of Shri Jagat Pati Joshi*, ed. C. Margabandhu, A. K. Sharma, and R. S. Bisht, vol. III, Delhi: Agam Kala Prakashan, 2002, pp. 696–701.
112 Samuel Beal, *Travels of Fah-Hian and Sung-Yun from China to India (400 A.D & 518 A.D)*, New Delhi: Low Price Publications, 2005, p. 67.
113 Thomas Watters, *On Yuan Chwang's Travels in India (A.D 629–645)*, New Delhi: Low Price Publications, pp. 335–336, reprint.
114 Cunningham, *Four Reports Made During the Years 1862–63–64–65*, pp. 272–274.
115 B. R. Mani, 'Excavation at Sankissa 1995–1996', *Puratattva*, 1995–1996, no. 26, pp. 43–44.
116 K. K. Sinha, *Excavation at Sravasti 1959*, Varanasi: BHU, 1959, pp. 5–8.
117 Takahiro Takahashi, Taizo Yamaoka, Fumitako Yoneda, and Akinori Uesugi, 'The Ancient City of Sravasti: Its Significance on the Urbanization of North India', *Puratattva*, no. 30, 1999–2000, pp. 74–76.
118 Yoshinori Aboshi, K. Sonada, F. Yoneda, and Akinori Uesugi, 'Excavation at Sahet Mahet 1986–1996', *East and West*, vol. 49, no. 1/4, 1999, pp. 119–173, pp. 122–24.
119 N. B. Sanyal, 'Predecessors of the Gahadavālas of Kanauj', *Journal and Proceedings of Asiatic Society of Bengal*, New Series, vol. XXI, 1925, pp. 103–107.
120 D. R. Sahni, 'Sahet-Mahet Grant of Govindachandra [Vikrama] Samvata 1186', *Epigraphia Indica*, vol. XI, pp. 20–22.
121 Aboshi, Sonada, Yoneda, and Uesugi, 'Excavation at Sahet Mahet 1986–1996', p. 126.
122 J. Takakasu, *A Record of Buddhist Religions as Practiced in India and Malay Archipelago (ad 671–695)*, Delhi: Munshiram Manoharlal, 1998, p. 86.
123 Anand Singh, *Nālandā A Glorious Past*, New Delhi: Primus Books, 2024, p. 391.
124 Anand Singh, 'Giriyaka Hilltop Buddhist Monastic Complex (Rajgir): Understanding Antiquarian Remains and Physical Spaces' *Ancient Asia*, 2021, vol. 12, no. 13, pp. 1–19, p. 17. https://doi.org/10.5334/aa.251.
125 Singh, *Nālandā A Glorious Past*, pp. 226–227.
126 Takakasu, *I-Tsing, A Record of the Buddhist Religion*, pp. 108–9.

5 Buddhist Outlook on Water and Environmental Concerns

Man's considered superiority over nature and planet Earth was first challenged in the 20th century when human societies realised that some of the major calamities and disasters like acid rain, melting of glaciers, rising of carbon gases, untimely droughts, inundations, and consequent water crises are a human-induced phenomenon. Veronica Strang says that

> *Earth may be the blue planet, but 93.3 percent of its water lives in its ocean and has varying levels of salinity. Fresh water is either found in rivers and lakes, contained in aquifers and underground basins, or held in ice. The largest reserve of fresh water on the planet is glaciers found in the high mountains of every continent except Australia. About a third of the world's human population, and a considerable percentage of its plants and animal species, depend on these for seasonal meltwater, thus the shrinkage of the glaciers due to climate change has significant hydrological implications.*[1]

The scientific innovations and discoveries failed to control Nature and her natural occurrences. A happy planet demands humans understand that all living and non-living beings are part of this natural world and that the Earth has limited resources that are not meant for unwarranted exploitation. The growing economic interdependence and globalisation have generated a state of affairs that accentuated closer social contact, an idea of global citizenship, and associations more than ever before. It rapidly increases the exchange of ideas, and new types of alliances are created, fostering socio-economic hegemony and hostility.[2] In the early world, natural resources like water were considered sacred and universal, but now monopolised, monetised, and desanctified. With the rising population, scarcity of water is felt in many regions. Pollution and industrial waste created the notion of troubled waters in many countries, which indicates that a catastrophe is imminent if civilisations do not understand the problem. Frequent droughts, untimely floods, storms, and paucity of drinking water in many regions left deep grudges on the human planet due to wanton industrialisation strategies, deforestation, and human colonisation. It requires urgent rectifications and solutions to

DOI: 10.4324/9781003601241-5

repose faith in conceptualising a happy planet. From the Anthropocene Age, the demographic shift due to the rising human population somehow affected the functioning of planet Earth. On one end, these anthropocentric endeavours led to the rise and progress of human civilisation to its pinnacle and on the other side, also caused phenomenal intensification of ecological disequilibrium, primarily visible in the form of cloudbursts, acid rain, floods, rising sea levels, etc. Their imposing influence is also perceptible on water, the most vital element to endure all kinds of biotic and abiotic elements on the earth. The existence of the entire surficial milieu of planet Earth, like the plants, animals, soil structure, and site, is entwined with water. It is a common understanding that water is in abundance and plenty. Water can reinvent and replenish itself, but its sources are not inexhaustible and subject to decay and end. Though living beings cannot survive without water, its sources and nature are still not understood holistically.[3] The misconstrued belief of the enormous and ever-existing water sources led to its misuse and unsympathetic actions. Water is misrepresented, contaminated, and misappropriated as the people of this century are more inclined to a market economy, profit-oriented production, and monopolised corporate culture that pays little attention to natural resources, water, and the environment.

Water is considered a natural, physical, and social element across cultures. In a social context, water is not a physical substance but a divine substance with its own cultural, metaphysical, and spiritual manifestations belonging to that particular culture. Natural and social sciences do not consider this traditional understanding of water appropriate.[4] The ancient belief in the sacredness of water was transformed during the Renaissance period, which led to the beginning of industrialisation. Scientific innovations have declared water to be a physical element. The market economy refigured it as a commodity that can be used and sold to make a profit. The volatile cravings to earn from every possible source have led man into unforgivable anthropogenic interventions to exploit and squeeze water resources. Recently, this has led to a vivacious subject of survival and the survival of the ecosystem, especially water in its natural state. Nature can recuperate itself, and it has been doing it over the ages to escape anthropogenic commotions and bring back splintered natural resources.[5] However, excessive human interventions make it impossible for Nature to recuperate itself. When the world started facing challenges like environmental degradation, water crisis, melting of glaciers, etc., many turned towards religious traditions and practices endorsing co-existence with Nature. The intrinsic values and teachings mentioned in different religious texts are vying for answers created due to unsolicited means of commercial production, industrialisation, demographic shift, urbanisation, and consequently the loss of water resources, flora, and fauna. Modern economics and the process of urbanisation have multifaceted and miscellaneous rules with special, unique, exceptional ethical standards based on profits and markets. It is believed that the maxim of maximisation of profit now governs Earth and the planet is to bear the pain that it has been sensing since the

Industrial Revolution. Mankind is doing enormous damage to their planet without thinking that water sources can be replenished, but it is not infinite. Some sections of policymakers and people are undermining the natural laws of ecosystems because they lack familiarity with planetary ecology. Their ignorance and indulgence in breaking the ecological codes will invite adversity and disaster.[6] It is pretty apprehensible and evident that Nature began responding as we grapple with untimely weather fluxes, increasing temperatures, inundations, droughts, and tsunamis. There is an urgent prerequisite to contemplating and resolving the new paradigms of rules and practices to check the prevalent material culture and restrain the further worsening of water sources.

Buddhism is an engaged religion that can guide humanity in developing discourses on water conservation and its sustainable usage. Buddhist monks and scholars are striving to create and produce the discourses that are originally inherent in the roots of the Buddha's teachings and are deeply relevant to contemporary issues like ecological and water crises. The Buddha has recommended the environmental use of water in many *Suttas* and, on some occasions, intervened to resolve disputes over water sharing. At the cosmological end, the Buddha was symbolised by a lotus and water. Buddhism has vast philosophical dimensions for four elements (water is one of them) and even deliberated how water from the great hills has a significant role in the universe's formation. Bodhi Bhikkhu says that issues like ecological deterioration, poverty, exponential population growth, etc., are fundamental existential disarticulations from the existing reality and distortions of the appropriate level of value. The teachings of the Buddha can save the world from the brink of darkness and gloom to the realm of bliss.[7] Buddhism has an ardent belief that imprecision of Nature, either intentionally or a failure to understand the laws of Nature, leads to the collapse and dysfunction of the whole ecological system. Maladministration of water sources and discerning water as merely a physical element are the essential causes of ecosystem disequilibrium.

Water Resources: Locating Environmental Issues

The marvels of planet Earth have also changed since its evolution, and contemporary geomorphic features have emerged due to different cosmic and natural occurrences over millions of years. Much before the arrival of our hominid ancestors on Earth, and because of those natural upheavals, the environmental and climatic conditions also kept changing.[8] Since their arrival on the scene, the hominids tried to understand Nature and harness the natural wealth for their growth, survival, and progress. With scientific progression, the *Homo Sapiens* were able to settle on the plain which led to the growth of their cultural settings and the genesis of civilisations. It also began to understand nature, coexist with it, and even conquer it for the enormous benefit of humans, and this human–nature conflict is still going on. However, the earliest civilisations seemed to be more environment friendly than the present-day human world, which developed the tendency of material progress lacking awareness and sentiments towards our natural surroundings. The recent devastating floods in major rivers of India and consequent inundations in the regions of Delhi, Himachal Pradesh, Uttarakhand, and some parts of South India are considered man-made by geoscientists. The floods and changing rain patterns in some parts of East Asia, encroaching heat waves in the European countries, and regular occurrences of thunderstorms on American coasts are also seen as a result of changing patterns of the environment because of rising temperature and pollution. The blatant rise of unplanned urbanisation and the making of roads, dams, hotels, and resorts led to heavy deforestation of the Indian hills, which made them vulnerable and fragile, as in some parts of the hills, the forests have completely disappeared. During monsoon season, heavy rains and the fast flow of water break the stone boulders and soil structure, which leads to disaster. Earlier, the dense forest in those regions could stop the fury of water, and trees could slow the flow of water coming from the melting of glaciers and heavy rain. Now denudation has become a pattern and an acceptable truth. In some parts, the seed rain of alien species led to forced climatic change, worsening the situation. The palynological studies and their outcome of some regions of India where floods became a regular feature suggest that deforestation is a major reason for the ongoing catastrophe. Though forest clearance started in the hoary past in the Holocene period, the changes were so gradual that they did not transform the geographical settings of the region and the major changes happened only on the side of the river banks and lakes. Even the expansion of agriculture over the past three millennia affected only the grasslands and not the forests. The significant ecological devastation began only in the 19th century after the Industrial Revolution.[9] The engrossing human intervention in natural ecosystems led to forest depletion, pollution of water and its sources, and increasing sea levels to abysmal proportions that submerged some parts of the mainland habituated by humans and animals in different parts of the world. The pastures' lands and meadows are now fast becoming deserts because of the increasing drainage rate and lack of enough water. The rivers fed by glaciers and rain gradually carry less water due to environmental changes. The rising beds of estuaries and rivers are slowing down the velocity of water currents, which leads to the transformation of shallow sea and fertile lowland into unproductive and miasmatic quagmires.[10] Human intention and desire demand a crucial shift from an obdurate material impertinence corroborated by the subjugation of Nature to a universal and fundamentally compassionate way to both biotic and abiotic species. It stresses liberal, interpersonal, and interactive co-existence through the justifiable usage of natural resources and their conservation. However, the prudence of such measures and actions also created confusion, arguments, and incongruities among different societies, communities, and nations. Many nations squeeze minerals and water resources enormously and dump toxic wastes produced by their industries in oceans, rivers, and hills outside their territory. It is contemptuous and harmful to humans and non-humans living in that area.[11] Acid rains, defiled groundwater, and water-borne diseases are creating havoc and killing millions of people across the world every year. The bulging rate of population, disproportionate use of technology, and excessive pesticides in the agriculture sector have increased land attrition, salinisation, and desertification. Hazardous wastes and toxic residues are polluting the entire world by atmospheric and oceanic circulations.[12] The quest for a feasible answer led to the understanding of religious tenets in a new way that can resolve the diverse gossamer of ecological issues and water crises. This new start is widely acknowledged as an engaged tradition. The commencement of the Buddhist-engaged tradition can be traced from an essay published by Lynn White in *Science* in 1967. The essay deliberated the role of Christian beliefs and their significance on the environment. He found that Christian ethics are more worldly and have diminutive care and respect for Nature. He argues:

> *What we do about ecology depends on our ideas of the man-nature relationship. More science and more technology are not going to get us out of the present ecological crisis until we find a new religion or rethink our old one. The beatniks, who are the basic revolutionaries of our time, show a sound instinct in their affinity for Zen Buddhism, which conceives of the man-nature relationship as very nearly the mirror image of the Christian view. Zen, however, is as deeply conditioned by Asian history as Christianity is by the experience of the West, and I am dubious of its viability among us.*[13]

The argument placed by Lynn White Jr. caused an eclectic uproar among experts in theological studies. James Gustafson refuted the hypothesis of Lynn White and projected five means of Christian ethics to deal with Nature: (a) despotic, (b) dominion, (c) stewardship, (d) subordination to, and (e) participation. He explains that between God and Nature, there are countless stages where humans engage over a period varying from despot to having dominion over Nature, but they still depend on it. Sometimes, man becomes a steward, and others treat nature as secondary to humans.[14] Pursuing the ecological debates, the scholars dealing with Buddhism advanced their hypotheses. Donald K. Swearer suggests that Buddhist concerns for the environment can be premeditated under five taxonomical classifications: (a) eco-apologists, (b) eco-critics, (c) eco-constructivists, (d) eco-ethicists, and (e) eco-contextualists.[15] To shape Buddhism into a distinct ecological position, a Vietnamese monk coined the term 'engaged Buddhism' with all characteristics of traditional Buddhism, including modern ideas and values connected to current issues. It was recoined as 'socially engaged Buddhism' under the stimulus inspiration of Thai monk Buddhadāsa Bhikkhu and other Buddhist scholars dealing with ecological issues.[16] Socially engaged Buddhism was a procedural renovation of Buddhist teachings and virtues to connect with the current socio-economic problems. It inspires extensive social movements to establish an ecumenical world of Buddhism for peaceful coexistence.[17] The stimulus for Buddhist engaged tradition was the Buddha, whose teachings facilitated the transformation of the socio-economic and political values of the world for a sustainable ecosystem.

Stanley Jeyaraja Tambiah and Juliane Schober found a set structure and form in developing engaged Southeast Asian movements. Tambiah mentions that Southeast Asian kingdoms were ruled by a galactic political organisation that incorporated cosmological, territorial, and politico-economic paradigms in a single framework. This web of structure can be compared with *maṅdala* evolved by Mahāyāna and Vajrayāna, encompassing a core (*maṅda*) and enfolding component (*la*). It comprises homogeneous and heterogeneous structures integrated into a nucleus and is well supported by the outer wings of Hindu-Buddhist religious ideas and practices.[18] The all-encompassing political vicissitudes in the post-World War era introduced political democracies worldwide in the 19th century. They transformed the polity of many Asian kingdoms/states and the functions of religious establishments across Asia. The decolonisation of Asia visualised the independence of many states in Asia. It also led to far-reaching changes in their political functioning, financial institutions, modern conveniences, and way of life. It also transformed the role, function, and rationale of the Buddhist Saṅgha in Southeast and East Asian states. Southeast Asia's political dynamics were transformed from traditional galactic states to modern nation-states. With sweeping political transformation, the relations between Saṅgha, the state, and society were redefined. It invigorated the centralisation of the Saṅgha, their increasing interactions with the state, the heightened role of the laity in ecclesiastical matters, and the development of a realistic religious worldview in the Southeast Asian Theravāda tradition. It also ushered in a new era of modern governance and economic reforms. The traditional economic edifices of many countries crumbled due to sweeping political and economic reforms to compete with the globalised economy. These structural and religious configurations have cemented the path for the recent Buddhist cataclysms in many Southeast Asian countries.[19] The monks started engaging themselves more in community development (*patana*) as well as in missionary activities. Many such monks are known for their welfare works and academic activities in Thailand and other Buddhist nations. Their actions earned accolades in many parts of the world, but muted voices of resentment were heard for the growing involvement of monks in the mundane activities. The conservative voices of the Saṅgha echoed apprehensions about their increasing role in society's material prosperity. Even the monks involved in community development are sometimes nick-named 'half-monks-half-laymen' and were cautioned to circumvent such social engagements.[20] Engaged Buddhism became a very popular social movement in all parts of the world. In its early phase, Thich Nhat Hanh, H.H. Dalai Lama, Sulaka Śivaraksa, A.T. Ariyaratne, Buddhadāsa

Bhikkhu, Joanna Macy, Paula Green, Stephan Batchelor, etc., were some of the leading monks and laities who laid a solid foundation.[21] One of the earliest such movements began in Thailand against market-oriented economic reforms of the Thai government in the 1980s, which caused extensive deforestation and unfavourably disturbed the natural networks of rivers and indigenous ecosystems.

Water and forests are the lifelines for sustenance of the Thai people and any hasty disturbance to alter the ecosystem could destroy their community and economic structure. Its effect could be seen in Thai society through floods and droughts. Chiang Mai, Bangkok, and other parts of Thailand faced unprecedented levels of flooding, and the cause was deforestation. The forest cover in Thailand receded from 72 percent of total land in 1938, and it diminished to less than 15 percent in 2003.[22] It was realised that eccentric strategies, guidelines, and actions are accountable for environmental damage. Some monks remonstrated against their government and implored that the teachings of the Buddha could impart real wisdom for the overall development of society and nation.[23] Their increasing concerns for the nation and people provided new opportunities for the Buddhist ecological movements. Their teachings highlighted the Buddha's teachings of impermanence to analyse the adversative influence of consumerism and the dependency of the economy on market forces. In 1974, Phra Dhammadilok discovered the possibilities of sustainable growth without damaging the resources of forests and other fiscal resources.[24] In 1988, Phrakhru Manas Natheepitak, Chief Abbot of Wat Bodharma in Northern Thailand, decided to ordain the trees to save them from destruction done by the contractors.[25] Seth Devere Clippard examines two diverse components betrothed by the heroes of Thai ecological movements: textual and engaged. They grabbed the help of the teachings of the Buddha and applied them to get rid of suffering caused by sudden economic transformations. The ecology movement can be designated as 'The monk is the signifier, the tree is signified, and the ordination is signification'.[26] Susan Darlington joined in one of the tree ordination ceremonies led by Phrakhru Pitak Nanthakhun and examined 'ecology monks' role. She observes how the monks are making people aware of depleting forests, water sources, and wildlife. They embraced the virtues of the sustainable connections between the Buddha and Nature.[27] In 1993, the monks planned a massive drive to resuscitate the Nan River, and their determination bore fruit. The monks also helped the native people raise local water bodies, excavate ponds for fisheries, and plant trees.[28] Thailand's 'ecology monk' drive could be equated to the *Chipko Andolana* (Chipko Movement) in the 1970s in the Garhwal Himalayas of India. It was intended to protect the trees and the native ecology of the region disrupted by organised deforestation. It was a *satyagraha* (peaceful protest) and non-violent resistance through the embracing or hugging of trees and surrounding them from human bodies to save them from being cut. It recalls the collective mobilisation of women of their forests and local ecology in one of the hill states of Uttarakhand, India.[29]

People of the Himalayan region of Uttarakhand, Himachal Pradesh, Sikkim, Leh-Ladakh, and other adjoining regions are great nature lovers. Since the past few decades, the whole climate-vulnerable region of the Himalayas has been facing severe environmental crises. Cloud bursts, flash floods, heavy and untimely rain, and floods are now common in the hills. They had some vague idea about the melting of glaciers, accumulation of water, and formation of the river systems. It is well known that the plate tectonics and collision of the Indian and Asian plates steered the upthrust of the Himalayas, Karakoram, Hindukush, and the Tibetan plateau. This phenomenon also started the formation of the subcontinent, the subcontinent's major rivers, and the arrival of the hominids. When the rivers changed their courses, the significant central alluvial land was carved out or swept away by the meandering courses of the rivers. Many species of plants and animals were born, perished, or altered due to the multiplicity of the geomorphic changes. In these unstable conditions, our hominid ancestors had little role in altering the water and forest ecology structure.[30] The critical factors in the Hindukush-Himalayan region are penetrating downpours, landslides, cloud outbursts, swift melting of glaciers, and failure of artificial structures such as dams and levees. The urban and rural regions of hill districts have been devastated by heavy rains and floods. The pollen analysis conducted at Parasaram Tal, Haigam, Khajjiar, and Rewalsar lakes in Himachal Pradesh and Naukuchiatal and Bhim Tal in Garhwal hills suggest that major denudation was a recent phenomenon.[31] Intense rainfall in the Himalayan regions of Kashmir and the hilly cold desert of Ladakh wreaked havoc and caused flooding in dozens of their districts. Cloudbursts are a common occurrence nowadays in the Himalayan region causing intense flooding and landslides. The cloudburst recently triggered flash floods and swamped some parts of the Leh district and it is seen that some remote areas were more damaged where the traditional flood channels and bunds were dismantled. Such happenings are due to unplanned development in hill regions, changing courses of rivers, and cutting of forests. These factors supplement the process of global warming and cause more evaporation of water. The increased evaporation is forming dense cumulonimbus clouds resulting in intense rainfall.[32] The people living in this region are Buddhists. They are conscious of the ecological restraints of the region and embrace locally accessible resources to endure themselves from the ancient past. Recent calamitous cyclical variations and adversity forced them to look back and stick to approaches that have grown over thousands of years of experiments. The Buddhist communities of Ladakh established an extremely well-organised and ecologically all-encompassing irrigation arrangement. Their indigenous exertion culminated in the immensely automated channelisation of water. They constructed canals with locally accessible resources, stone pieces, and boulders. The accrued snow started melting as *kang-chu* (ice water) in the rivulets merging at some pivotal points and forming a *togpo* (stream). The *togpo* flows in the valley touching the settlement clusters and the villages and is associated with the *ma-yur* (mother channel). In the growth of this traditional design of irrigation, the Buddhist way of life leads the entire procedure. The Lama of the village appeases the spirits of water, earth, hills, and insects before the commencement of the sowing season. The hymns devoted to Manjuśri are recited by the monks to pray to him and the other Bodhisattvas for high yield. This is an inimitable way of man's coexistence with frozen water and snow.[33] However, their accustomed technique of agriculture and irrigation became endangered because of the fast melting of glaciers before the spring season, that is, their agricultural period.

Several people in Ladakh believe that the increasing footfalls of tourists as well as the shifting climate and weather conditions are accountable for this catastrophe. The hill people of Leh-Ladakh evolved a native mechanism to face the challenges of weather fluctuations. They went back to their traditional way of sustaining themselves but also protested against the rising unplanned hotel industry, the arrival of a puzzling number of tourists, the extension of orchards in eco-sensitive zones, and unplanned industrialisation in the eco-fragile regions. It has been noticed that the glaciers surrounded by highways and encroached by human actions are melting at a relatively faster degree than the glaciers situated in distant regions. Ladakh has had a massive number of tourist footfalls in recent years because of its mesmerising topography, high-altitude lakes, and mesmerising Buddhist culture. Global warming undeniably has a key part to play in climate change, but uncleanness due to the influx of several thousand tourists, a plethora of human waste, and increasing carbon releases produced by waves of humans are equally responsible for the harm. The incessant intrusion of the huge number of tourists between the months of April to August every year makes the region more fragile and eco-sensitive. The high-altitude and lower level of oxygen lead to a high level of drinking water as well as the use of toilets by the tourists. Most of the natives of Ladakh still practice the traditional method of toilets that consumes less water than the Western style. They are now imbibing with new kinds of toilets in which they are using manure in place of water. The accomplishment of this kind of procedure will save thousands of litres of water used in Western types of toilets. The packaged water is another kind of threat to the local environment and millions of unfilled plastic bottles used in the tourist spots in Leh district were hand-picked and dumped in the Nubra Valley of Ladakh. The rising stimulus of tourism also imbalanced the urban–rural continuum as a large number of people migrated to Leh district to support the tourism industry, triggering inequity in the traditional style of agriculture.

Ladakh's freshwater sources depend on glacial snowmelt. With the noticeable effects of climate change, receding glacial lakes and streams created difficulties for the natives. Due to severe climatic upheavals and rising stages of pollution, Ladakh is now facing a multi-dimensional water crisis like freshwater shortage, water pollution, lack of enough water for irrigation, getting enough share of water in the villages, etc.[34] Buddhist associations, monks, and ecologically conscious people of Leh-Ladakh are coming forward to save the local ecology. Every year in the winter season women, men, children, working professionals, commoners, etc., come and pray to the Bodhisattvas Amitābha, Manjuśri, Avalokiteśvara, and other local spirits and unfurl the Buddhist flags even at -30 to 40 degrees Celsius in their villages, schools, junctions, etc., to create awareness and responsiveness about the sacredness of water.

They have come up with Artificial Ice Reservoirs, also known as artificial glaciers (AG) which are structures designed to store water in the form of ice. These are created through the process of glacier seeding. Sonam Wangchuk and Chewang Norphel, the Buddhist conservationists and ecologists of Leh districts highlighted that if such thoughtless and inconsiderate development lingers in Ladakh and other parts of the Himalayan region, the glaciers will be extinct soon and India will face anxious levels of water shortage. They, with the help of the Buddhist monasteries of Ladakh, pioneered ice reservoirs near village streams that are fed by perennial natural springs or glacial streams. There are three main types of AGs: Basin Type Artificial Glacier (bAG) is simpler and requires low maintenance. Here, a basin is constructed between the village and the glacier, directing water flow into it. The bAG acts as a pond during winter, allowing for activities like ice skating and hockey. In spring, water is released by lifting a gate valve for distribution among the Buddhist Communities; the second one is a Cascade Type Artificial Glacier (cAG), which involves building retention walls at regular intervals across the stream, forming a step-like glacier. It is a simple and effective method if the stream is not exposed to direct sunlight in winter; and the third is a Diversion Type Artificial Glacier (dAG), the most popular form, involving diverting water from the main stream through a narrow canal to a shaded valley. The water freezes in the valley through a series of walls, similar to cAG. This technique is suitable when the stream is exposed to sunlight in winter but requires more investment, labour, and maintenance. The Artificial Icefall Glacier (AIG) was introduced by His Holiness Chetsang Rinpoche of Lamayuru Monastery in 2016 and has since been replicated across the western Ladakh Sham Valley region by his followers. AIG is characterised by its low-cost and low-tech features, requiring only a water source, a pipe, and a shaded cliff face. This simplicity allows for easy management and construction without external intervention. The Ice Valley Prototype Project has implemented AIG in two locations around Leh. Villages like Ursi have been constructing AIGs for the past three years, using meltwater to irrigate fields. Since its inception in 2018, Ursi Village has managed to irrigate 100 canals of land, a significant increase from previous years plagued by water shortages.[35] The Ice Stupa (IS) Project, conceived by Sonam Wangchuk, addresses the limitations of traditional AGs by freezing under-utilised water during winter months for later use in agriculture during the spring sowing period. The project's prototype, constructed at SEMCOL Alternate School in Leh, was successful and lasted until mid-May. Encouraged by this success, SECMOL collaborated with Phyang Monastery to implement a full-scale structure in Phyang village in Ladakh during the winter of 2014–2015. This involved freezing stream water vertically to create ice towers or cones ranging from 30 to 50 metres in height. The vertical stūpa design overcomes challenges associated with horizontal AGs, such as specific site conditions, high-altitude requirements, constant maintenance, and north-facing valleys, making it a more versatile solution for water conservation in arid regions like Ladakh. The Ice Stūpa Project has significantly expanded its efforts through annual ice stūpa competitions and training sessions, involving over 50 villages and 400 stakeholders in building ice stupas across Ladakh over the past three years. This initiative has greatly contributed to addressing water scarcity and supporting communities in regions like Phyang, Kulum, and Gangles villages (Figure 5.1).[36]

These *stūpas* are a novel method of water harvesting, a sustainable and ecological device to create AGs to fulfil the requirements of locals, especially the agricultural communities, to mitigate water scarcity and greening the arid region. They develop a conical structure that resembles the *stūpas*. The ice structure was developed by taking the natural flow of water from the hills in the winter season. The sub-zero temperatures help the flowing water to freeze in a conical shape and it looks like the ice domes with hanging ice strings. The dried branches of trees and fodders are messed up and put at the centre. Sea buckthorn (*Hippophae rhamnoides*) plants and leaves are also used to shape the structure of ice. The conical shape facilitates the exposure of a minimum surface area to the external temperature in the months of summer, enabling the slow melting of ice and preservation of artificial ice reservoir for

Figure 5.1 Ice Stūpa (Leh-Ladakh, India), Courtesy: Stanzin Mingur.

a long time. Because of the overwhelming Buddhist influence in the region, the structure is named the '*ice stūpa*'.[37] The '*ice stūpas*' of Ladakh are a locally devised climate-adaptive mechanism promoting water husbandry in the arid region and is a ground-breaking answer that high-altitude Himalayan Buddhist communities have developed as a response to mounting ecological disturbances jeopardising their life and topography.

The prevailing archetypes have already established their capability to accumulate a large quantity of water, paving the way for sustainability in field irrigation and domestic use. The experiment can be acknowledged as a great success.[38] Ecological shifts and local microclimate turbulences played catalytic roles in diminishing the accessibility of glacial water due to the reduction of glacier mass. The creation of these human-devised ice masses will facilitate an ardent mechanism to deliver a valuable substitute, empowering farmers to store ice during the winter months and use the meltwater in the spring, a season of water scarcity in Ladakh. It cannot stop the receding and melting of glaciers but unquestionably serves the people to acclimatise to climate change.[39] The Buddhist associations of Ladakh and other such agencies organise competitions every year across the villages and suburbs to make bigger '*ice stūpas*' and between November to March, one can find hundreds of such '*ice stūpas*' encircled by Buddhist flags in all parts of Ladakh. The locals living in Leh informed me that it was a small beginning a few years back and now taking the shape of a global movement to save glaciers and water. Its scope and popularity are now extended to the hills of Nepal where Buddhist communities started making '*ice stūpas*' in the winter. The kernel of such kind of environmentalism is inherent in their Buddhist training and culture that evolved and developed in these hilly terrains. Since their childhood, the children are trained not to pollute water springs. It is believed that these springs and water sources are home to benevolent spirits. Such a wonderful attempt to restore and preserve the Himalayan ecology by their people can be appropriately termed 'ice stūpa ecology'.

The problem is not limited to oceans, rivers, and hills but sources of potable water are also under constrictions. A substantial number of the population lacks enough freshwater for drinking and cleanliness.[40] Even the mechanised form of supply of drinking water in urban spaces is questioned by some social scientists. They consider that water in its natural form is the purest. Earth was formed 4.6 billion years ago when magma released hydrogen and it got transformed into steam due to atmospheric activities, then evaporated, and formed rain. When it fell on the earth, the great oceans were formed. Water that comes from the sky on the earth became part of groundwater over a long time. Joan S. Davis says these kinds of waters which fully became part of groundwater over the years are 'wise water' and are the healthiest. And the water that has recently rained is 'juvenile water'. She says that water in its natural form is the healthiest for humans but the quality of water is quite deplorable that is coming through long pipes through mechanised efforts. In the pipes, the water is exposed to a high pressure and high flow mechanism which breaks its mineral contents and compromises its quality. It is an urgent need that humans start showing more respect to water as it is their life saviour.[41] The traditional way of harnessing water developed by the traditional societies in the early phase of our civilisation deserves more consideration than it is found from the modern hydraulic mechanism. They developed better consciousness and well-conceived planning to harness the water without disturbing the pattern of precipitation and natural water storage bodies.

The prevalent form of irrigation and raising of dams also came into the domain of debate when the water crisis emerged. When big dams are built for large-scale irrigation and hydroelectricity projects, they facilitate irrigation over a vast area but also engulf and inundate a large chunk of land in the region. Sometimes several villages are submerged and a large section of the population is displaced. These dams change the natural ecology of the river and the ecology of the habitation of the natives. Though it has economic benefits but also has demographic and economic deprivation. The changing geography creates a gap in the social and cultural fabric from which the local people were knitted over the centuries and generations. With the growing population and innovation of technology, large dams were built across the globe in the 19th and 20th centuries to command and redirect the large sources of water for the generation of hydroelectricity, expansion of agriculture, and irrigation. The dams were considered the symbol of prosperity and development. Jawahar Lal Nehru called the large dams 'temples of modernity'. Soon due to ecological fragility, the breaking of local socio-cultural ecology, deteriorating soil structure, and change in the geomorphology of the region, the making of large dams was criticised.[42] Though ancient and medieval civilisations and cultures also built large dams till 1949, about 5,000 large dams were raised worldwide, three-quarters of these are in industrialised nations. By the end of the 20th century, 45,000 large dams were built in over 140 countries.[43] It is not only large dams but the making of large chains of canals is also questioned as they are destroying the natural ecology of the region. Worster argues that man-made canals do not have their ecosystem. It is a directed mechanism of humans against the environmental virtues of the region to divert a definite volume of water through a concerted effort to make money. He says that canals along with many others like it have chain-lock fences running on either side, sealing the ditch off from stray dogs, children, fishermen, and human life. It is not only large dams but the making of large chains of canals is also criticised as they are destroying the natural ecology of the region. Worster argues that man-made canals do not have their ecosystem. It is a directed mechanism of humans against the environmental virtues of the region to divert a definite volume of water through a concerted effort to make money. He says

> *...Canal along many others like it, tall chain-lock fences run on either side, sealing the ditch off from stray dogs, children, fishermen, solitary thinkers, lovers, swimmers, loping hungry coyotes, migrating turtles, indeed from all of nature and of human life...*[44]

However, the making of large dams and canals also extensively changed the nature of the agrarian economy. It constitutes 20–40 percent of irrigation across the world and is a major contributor to the large-scale production of hydroelectricity which constitutes almost 20 percent of total electricity production. Some of the projects also helped in smooth navigation and avoidance of measuring natural disasters like floods and droughts. However, it is also true that the making of large water reservoirs displaced millions of people and changed the natural ecosystem.[45] The urgent need to settle the question of whether or not to make large dams should be examined by choosing the strategy of the *majjhima-magga*.

The management of water and its resources is changing fast around the globe. People of the arid region are trying to find new water sources to meet the ever-growing demand. This changing notion ushers in efforts to fulfil the bare human need, and the concerned governments added this agenda to their policies. Efforts have been made to meet the demand without any far-reaching change or large water transfer from one region to another.[46] Water's historical, cultural, and philosophical milieus differ from the approach of modern hydrological parameters and practices. In early socio-cultural strata, water's sacred and corporeal nature is more emphasised than the contemporary perspective of water as an economic commodity. Since the 20th century onwards, several hydrological projects have been initiated in all parts of the world that have compromised the natural flow of rivers. The water resources of the rivers of all continents were affected by this mechanised water management by raising hydrological projects, merging and diverting river courses, making large canals, etc. Whether a colonial authority, states, nations, political elites, or the dominant communities of the region whosoever engaged in new kinds of experiments and intervention to bring control over water, failed to understand the natural momentum of rivers. It is hypothesised that this new way of hydraulic management sought to correct past wrongs and it will protect the regions from frequent deluges. The arid areas will be fertilised with the new irrigation system and water wasting will be minimised. These measures also have some harmful effects, and sometimes unimaginative thoughts and plans can disturb the existing ecosystem of rivers.[47] The water crisis, a sensitive issue, is crossing a dangerous line and threatening human survival in some parts of the world. The situation is distressing as more than one billion people drink water polluted with human waste, and innumerable millions grapple with insufficient supplies of water. Their crops will suffer due to a lack of water. Population growth and climate change are collaborating to aggravate the situation. In the next few decades, the normal share of water per person will decrease by a third. The rising exigencies for water by humans impend natural ecosystems to an amount that may bring nations into tensions, skirmishes, and fear of war. The felonious intentions are challenging the concerns of policymakers and diplomats to the edge. The world still has not engaged this problem seriously and in the future, it will be problematic for hydrologists, meteorologists, epidemiologists, and legislators to get an appropriate resolution. To date, the efforts made by the people and governments are not convincing.[48] Linton claims that water as a sacred and natural element is ignored and is now treated as a commodity. This concept of water is 'modern water'. Scholars, scientists, and intellectuals have dedicated less time to the scarcity and the apparent water crisis. They are more vulnerable to climate change, deforestation, melting of glaciers, etc. But people have to think that only three percent of total water is fresh, and it's small quantities, rest in lakes and rivers, and the rest is either ice-covered or underground. Ninety-seven percent of the total water is in the oceans. Freshwater is limited and efforts to control and monopolise the sources will be mistaken. Water must be preserved as precious and a sustainable method to sustain it is imperative.[49] Out of the planet's surface, 71 percent of water is on earth and only 2.5 percent is freshwater. Of the entire water consumption, 70 percent goes to irrigation, 10 percent to domestic use, and 20 percent to hydropower. The world is besieged with the water gap, and it is contemplated that 1.3 billion people will have less access to safe drinking water in the coming years and that 2.6 billion will face a shortage of proper sanitation. It is vital to set up an appropriate water policy and device with the support of governments and private enterprises.[50] Finding a resolute response to deliver the least required water to all is significant. Two billion people still lack safe water, and one billion do not have appropriate potable water.[51] Governments and legislators worldwide must leave control, domination, and regional divide to seek a sustainable method.

Eco-Buddhist Discourses

Buddhism was born in the surroundings when the Indian subcontinent was struggling with many socio-economic problems. The Buddha discovered the root cause of these subjects and provided agreeable elucidations. He was conscious that man is unceasingly reliant upon Nature and that unsolicited meddling to endanger the ecological equilibrium will frowningly interrupt the people and society at large. When such occurrences started happening in the 20th century, the Buddhist notion of interdependence was believed as the best means to cope with such a crisis. In the economically cohesive ecosphere, Buddhist teachings of interdependence show that all sections of society should have equivalent access to natural resources, and the socially oriented mode of governance is more apposite and suitable. The Buddha always supports that the actual source must be acknowledged, which causes conflicts and suffering, because the elimination of the reason will be the finest method to reach a feasible solution to the problems.[52] In this instantaneous stage of globalisation, industrial junk like nuclear wastes, chemicals, pesticides, etc. are thrown into the water of the rivers and oceans. In modern urban spaces, most sewers are connected to rivers, and in the absence of efficient sewage treatment plants, they continue to pollute the water. The majority of the rivers of India are polluted due to the incessant inflow of industrial wastes and sewage drains. The people need astute planning to sort out these issues and their approach should be sustainable and enduring.

Economic development should be planned and implemented with a more significant viewpoint to take care of the welfare of the human world and the issues that are causing problems to them.[53] E.F. Schumacher, who has advanced the theory of 'Buddhist Economics', articulates that the existing mode of economic production and flow is grounded on the unstable nature of consumption rate and manufacturing enterprises. Nowadays, acquiring wealth is the highest form of asset, and wealth is measured to be the most authoritative means of society. It necessitates rethinking because the obsession with wealth now and then injures the self and Nature. He suggests the Buddhist view of restraint and self-control as the best substitute to handle such problems.[54] The teachings of the Buddha always inspire calmness, rational thinking, and harmonious living. Mindfulness and unceasing commemoration of the teachings of the Buddha will augment empathy and egalitarian qualities. Buddhist tenets constantly assure people to espouse milieus founded on shared values, a justifiable mode of production rather than rapacious profit market economies.[55] The increasing commercial interconnections, augmented socio-cultural exchanges, speedy progression in information technology, and geopolitical contests are compelling the human world to be socially more unified and shaping the planet more resolutely in a single global network. The fast-growing internationalisation of society and economy also triggers contentions and struggles driven by spatial, financial, and ethnic diversities and disparities. The expropriation of resources now becomes the order of the day, putting substantial pressure on human lives. Every country is demanding its exclusive water zone. The distribution and sharing of water and its resources are causing geopolitical discontent and conflict. The natural landscape of the rivers is now divided among the many nations, and each one is trying to control the water. At this moment, the Dhamma becomes relevant, which speaks that greediness is the worst of diseases, contentment is the utmost wealth, and trust is the finest of relationships. It enhances the embryonic growth of a shared culture that can help redesign contemporary social, economic, and administrative organisations. It enunciates the most trustworthy principle of eliminating the human sufferings known to mankind. The dearth of elementary necessities and infrastructure amenities in any region or nation generates malfunctions and corrupt forms of governance. Gross disagreement with underdevelopment, over-indulgence, and avariciousness by the advantaged classes also produces chaos in society, and both processes, that is, the under-consumption by the masses and the over-consumption by the advantaged classes, go instantaneously.[56] The Buddhist ethics of equivalence and non-discrimination constantly precede equality and social justice circulation across countries and continents. It recognises the source of the problem and the factors causing it, whether an individual or a regime, and gives a viable answer that is advantageous and acceptable to all irrespective of nation, religion, or race. They can co-exist and harness the natural resources more harmoniously.

Buddhism talks about non-attachment to self and mundane matters, but it also recommends self-care, consideration, and kindness for other sentient and non-sentient beings. Non-attachment and callousness cannot be treated on the same platform. The Buddhist view of non-attachment does not prevent anyone from undertaking a constructive approach to others.[57] Since its genesis, Buddhism proposed concepts about synchronisation and friendliness with Nature. It deliberates their position, usage, and coexistence with humans and contemplates their beginning and growth. The knowledge looking for the previous acquaintances of humans with biodiversity was perhaps to deliver a more lasting relationship between man and Nature. Greed is considered one of the most rancorous reasons causing irredeemable environmental impairment. The Buddha ardently scrutinised this factor and acknowledged it as a key source for the decay of civilisations. The *Cakkavatī Sihanāda Sutta* illustrates how unwarranted greed and unscrupulous policies can lead to the downfall and extinction of a civilisation. The story starts with a righteous person of the past who had a very long life up to 80,000 years and was unceasingly gifted with charm, wealth, health, and longevity. When the people began deviating from their Dhammic path and indulging in inconsequential matters, their life spans diminished and their charm, wealth, and happiness deteriorated consistently. In the future, as morals continue to degenerate, human life will also start to shorten to the point where the typical life span becomes 10 years.[58] The *Sutta* says that the glorious attributes of a Cakkavatī, his righteousness and his intention to rule his kingdom according to the Dhamma, can endure his kingdom and the universe. The *Cakkavatī Sihanāda Sutta* talks about two significant messages: hedonism always leads to downfall and people must evade greed as it will denigrate and finally destroy the planet. It also highlights the role of upright governance in upholding the political, social, and economic equipoise. Water is a natural resource on which the whole planet subsists, and it is the concern of the states and the people to prevent themselves from wasting and destroying water sources and quality. The *Sutta* again apprises that at the time of dire need, the Buddha was born as Dhamma Cakkavatī Metteya (Maitreya) to save human civilisation from extinction. He bestowed blessings on the Cakkavatī Saṅkha, the ideal king to save the earth and the people from unending peril with his benevolent rules and good governance as the duty of the perfect king is to encourage and inspire the Dhamma in the life of the familiar people. The Cakkavatī is the role model as well as the cultural epitome of the society and his act of generosity is prevalent and superimposing. The Buddha is also a Cakkavatī and his teaching has manifold effects on the cosmos.[59] S. J. Tambiah says that the *Agañña Sutta* displays the way and procedures for establishing social order and political statecraft. The Buddha favours the notion of the Dhamma over the *Artha* (economy) in the governance domain. The Buddhist literature informs that all moral and cosmic laws are entrenched in the *Buddhavacana* and incorporates the path to liberation and the code of kingship. The *Buddhavacana* shows the path to the kings and guides how to take inspiration from the Dhamma in the conduct of temporal affairs. The law-framing agencies should prioritise the universal moral law, and nations must accord moral epitomes embedded in the Dhamma. State jurisprudence should not be considered supreme unless inspired by the Dhamma, the universal moral law that also encompasses the law of kingship.[60] The viewpoint of the Buddha was to find egalitarian societal standards that could cut across communal obstinacies and strangeness grounded on race or religion. He demonstrates that lineage or race does not form man's destiny. It does not facilitate good or bad deeds, good looks or wealth.[61] Buddhist tenets give us new forms of intellectual thoughts about the ecological crisis, including the shortage of water, and these wisdoms are a vivacious strength to deal with these issues more generously. It can function as a more realistic action to comprehend and appreciate the true nature of global competitiveness and contests to accomplish the wide-reaching commitment for water, vis a vis, makes a sympathetic working milieu to preserve its sources.

Many natural disturbances, such as floods, landslides, storms, and other climatic changes, like depletion of biodiversity, loss of aquifers, and drying of the rivers, can be linked to our failure to understand the water ecology. With the growing population, there is a massive increase in fetching water from different sources to meet demand and supply. Large dams, canals, sluices, and reservoirs were made to supply potable water and irrigation. At this juncture, serving the needs of the people is a priority but nurturing the water for the future is also a great worry. It is a predisposition of Buddhism that when people are suffering, the Buddha will come to save them and the path will be the *majjhima magga* (middle path). The middle way is a path that leads to the cessation of a negative attitude. The Buddha disapproves of the wrong or extreme paths the monks and laypeople followed. He tells the monks:

> *Monks, these extremes should not be followed by one who has gone forth into homelessness. What two? The indulgence in sensual happiness in sense pleasures, which is inferior, vulgar, low, ignoble, and hollow; and the pursuit of self-mortification, which is painful, ignoble, unbeneficial. The Tathāgata has awakened to the Middle Way, which does not get caught up in either of these extremes, which gives rise to vision, which gives rise to knowledge, which leads to peace, to direct knowledge, to enlightenment, to Nibbāna*[62]

This path illustrates an all-inclusive code of conduct for human life connecting it to the natural laws and favourable to the highlighting of good virtues in this life and the afterworld. It can be the most inimitable manoeuvre unanimously applicable to save the world from disproportionate consumerism and a monopolised market system. Much discussion exists about how Buddhist wisdom can be betrothed to scrutinise the problems that only appeared in the 20th century. Buddhism originated 2,500 years ago, and some of the issues, such as water depletion, the drying up of rivers, the paucity of water in certain areas, and tsunamis, are current issues. David Loy says that consistency and rationality were ensured between the teachings of the Buddha, which first originated in the 6th century BCE, and current issues developed and continued with the advancement of modern scientific knowledge. The Buddha is omniscient and sees everything past, present, and future. Still, it is difficult for him to know about the contemporary sciences, anatomical studies, atomic structure, or the laws of gravity. Religions are supportive of gradually changing the outlook of society. The canons, instructions, and practices stimulate the character and behaviour to transform and improve. The wisdom of the Buddha is meant to support mortal life and spiritual awakening.[63] These teachings had a captivating effect on society in the ancient past and are still relevant to put forward the kernel of good governance and welfare economics. The rising human population and shrinking natural capital are putting enormous pressure on the planet's ecological drapery, which has limited resources to sustain the world. It is acknowledged that the world will soon start facing a shortage of natural resources such as oil and natural gas, which will be exhausted in the coming years. Human and industrial waste is increasing water pollution and threatening the sink capacities of the earth. Natural resources like topsoil, freshwater, fishing grounds, and regional biodiversity are diminishing. In Buddhism, it is known as social dukkha, which is inadvertent or spontaneous and has corrective measures.[64] Buddhism gives the distinct prospect to inspect and correct the socially and economically made problems as everything that arise and exist in this world is impermanent, so the culminating water crisis can also be fixed. The teachings of the Buddha reassure people about getting rid of sins like greed and the accumulation of material possessions. By adhering to his ideals, the people will be more driven to share the water resources rather than their misuse.

The responsiveness and compassion described in the Buddha's teachings in a non-metaphysical mode lead to a shared understanding of societal paradigms. Stimulating the early Buddhist canons to form social ethics and comprehend water's true nature and significance in our lives inculcates a new methodological approach. The *paticcasamuppāda* can be used to observe the source of the water crisis and its regional implications. As the cause of the crisis is not universally applicable, the doctrine of *paticcasamuppāda* will help to scrutinise the issue at the micro level. It will show the human's actual place in the universe. In some regions of the world, these experiments have already happened as seen in the movements like the 'ecology monks' and '*ice stūpa*'. Here the teaching of interdependence has been reinvented to learn and offer answers for ecological glitches like water crisis. As we started seeing the Dhamma in a non-religious framework, this helped us learn Nature's actual characteristics. The Buddha incorporated the middle path by evading extremes advocated by believers in other Śramanic and Vedic sects. In the *Paccaya Sutta*[65], the Buddha comprehends the four qualities of *paticcasamuppāda*: objectivity (*tathata*), necessity (*avitathata*), invariability (*anannathata*), and conditionality *(idappaccayata*). The first describes the objectivity of the actual causal relationship. The conditions like inevitability and consistency highlight the absence of any exceptions. A resolute set of settings leads to a firm result and cannot be different. If circumstances are not adequately recognised, then the resultant happening identified in this phenomenal world cannot be inspected correctly. Accidental incidences or occurrences without any causal form are just ignorance of the causal process. The reason for the water crisis could be observed by eradicating the ignorance of its cause and acting upon its refinement. If one confesses that there is no accidental cause for such corrosion, it will be easy to determine the actual cause. The Buddha says when an erudite man identifies the condition, recognises the source of the condition, and distinguishes the ending of the condition, he finally learns the path leading to the cessation of the condition. He becomes a noble one who has learned the true nature of Dhamma. The noble beings of this mortal world have to identify the real cause that is creating the problem. The Dhammic view will find a clear differentiation between need and unequal expropriation. It will be a genuine Dhamma and the persons having such virtues are noble ones with piercing and permeating wisdom, one who stands tall before the door to the *Arahanthood*. The *paticcasamuppāda* can be combined into *kuśala kamma* or moral action and during its application in worldly engagements, it should not be taken in a metaphysical sense but as an assumption to know that nothing happens without a cause, yet no outcome is predetermined because causes are multiple and mutually intervening, giving an opportunity to new as well as existing. It is essential to Buddhist beliefs and can be a new path to engage issues like environmental deterioration and shortage of water. Joanna R. Macy says that ethics is entrenched in this new understanding of interdependence, as in the outcome of Buddhist views of *anatta* and *karma*. Consequently, it divulges shared dynamics between personal and social transformation as articulated in Buddhist canons. She endorses that:

> *All factors, psychic and physical, subsist in a web of mutual causal interaction with no element or essence held to be immutable or autonomous. Our suffering is caused by the interplay of these factors and particularly by delusion, aversion, and craving that arise from our misapprehension of them. We fabricate our bondage by hypostatizing and clinging to what is by nature contingent and transient. Being so caused, our suffering is not endemic. It can cease, and the causal play reversed. This is achieved by seeing the true nature of the phenomenon, their dependent co-arising. This seeing, which amounts to a reorganization of personality, is made possible by cleansing the perception (through meditation) and by moral conduct.*[66]

Frank Reynolds supports the same view and says that the *paticcasamuppāda* is crucial in understanding the early Buddhist view of *kamma*. He found that Buddhists rely on the Dhammic virtues while dealing with cosmological viewpoints.[67] The notion embedded in the doctrine of *paticcasamuppāda* shows the series of causal connections related to actions and their effects. It lacks any form of determinism but offers the groundwork for a law of just reprisal (*kamma*).[68] A causal relationship has been defined mainly concerning human personality. In the same way, environmental deterioration and water crises could be compared with the suffering faced by the human world, as worldly pleasures appear to be such only to those who are short-sighted and conceited. Its actual cause has not been recognised; thus, human life is entire of misery as nothing is unconditional and every happening depends on some conditions. As the existence of every eventuality depends on some reason, there must be something from which human misery comes into existence. The *paticcasamuppāda* not only explains the arising of *dukkha* but also provides remedies. It stimulates one to realise individuality and *samsara*. It demonstrates the correlation between action (*kamma*) and consequences (*vipāka*) just as a human being is dependent on plants, animals, and water, and deviating from values of coexistence with them, only invites disasters. Even the mind is dependent because the beginning of thoughts relies on the same spectrum derived from the external world of objects and persons. Each and all tangible and intangible things are constantly changing, transformed, subject to ageing, decay, and finally perishing. This phenomenon was universally applicable in different times and spaces and applicable to both sentient and insentient beings. Bodhi Bhikkhu says:

> *But I do believe that the religious crisis of modern humanity is intimately connected to these diverse social and political tragedies at many levels. Some of these levels, I would add, lie far beyond the range of rational comprehension and defy analysis in terms of linear causality. I would see the connection as that of co-arisen manifestations of a corrosive sickness in the human soul —the sickness of selfishness and craving — or as karmic backlashes of the three root defilements pinpointed by Buddhism — greed, hatred, and delusion — which have become so rampant today. I therefore think that any hopes we may cherish towards healing our community, our planet, and our world must involve us in a deep-level process of healing ourselves. And since this healing, in my view, can only be successfully accomplished by re-orienting our lives towards the Ultimate Reality and Supreme Good, the process of healing necessarily takes on a religious dimension.*[69]

Ignorance (*avijja*) leads to evil or unwholesome behaviour and when it is completely wiped out, it culminates into enlightenment. The Buddha says that *samma ditthi* can eliminate *avijjā* (ignorance) and *taṅha* (craving). David J. Kalupahana says that *avijjā* always plays a significant role in developing the dispositions (*saṅkhāra*) that establish the nature of man's behaviour. The beginning of the consciousness (*viňňana*) also depends on the nature of dispositions. Consciousness is the source that sets the nature of the new psychological aggregate (*nāmarūpa*), the role played by the dispositions in influencing the life after death is emphasised. Dispositions are consequently responsible for the nature of one's behaviour (*kamma*) and one's future birth or rebirth (*punnbhava)*. The course of rebirth is elucidated as combining two factors: consciousness (*viññana*) and the psychological aggregates (*nāmrūpa*). The psychological aggregate is the foetus developed in the mother's womb (*gabba*) and it signifies the commencement of a new life span.[70] Desire always leads to *taṅha* (craving). Understandably, craving for material affluence preludes unsolicited actions. Such actions are central to possessions that generate avarice, which further causes strife, anxiety, conflict, and worries.

The *paticcasamuppada* stimulates one to realise individuality and *saṁsāra*. It demonstrates the correlation between action (*kamma*) and consequences (*vipaka*) just as a human being is dependent on plants, animals, and water, and deviating from values of coexistence with them only invites disasters. Even the mind is dependent because the beginning of thoughts relies on the same spectrum derived from the external world of objects and persons. The *Kośambiya Sutta* shows the method the water crisis can be resolved. While reflecting on the unfortunate occurrence in the Ghositārama Vihāra, the Buddha says that it is not requisite to become enlightened or get conversant with the doctrine of *paticcasamuppāda. Still, it* is appropriate to understand the true nature of the problems that accentuate its elimination.[71] If one fails to understand the true nature of problems, then the real cause and effect relation would not be identified. It will lead to the knowledge of a fractured methodology that may cause catastrophic results or failure. In the life of the lay people, the *paticcasamuppāda* has not been used to explicate the arising and ceasing of suffering. Still, the causal effects were taken into account to solve their routine affairs. It could be applied to contemporary issues like the water crisis. The Buddhist canons suggest suffering as an individual's feeling but more largely, it is acknowledged that an individual is a part of society. So, the suffering of an individual might be the suffering of the entire society. The Buddha prohibited the metaphysical and absolutist nature of suffering saying that individuals create it. Both reality and humans are subject to some extreme attributes. These attributes are *tilakkhana*, that is, *anicca* (impermanence), *dukkha* (dissatisfaction), and *anatta* (no substantiality). Buddhism believes in impermanence, that is *sabbe saṅkhāra anicca* (all conditioned phenomena are impermanent).[72] Here *saṅkhāra* comprises reality demonstrating that all animate and inanimate beings share certain fundamental attributes and dependence. The Buddha always advises his followers to distinguish between the mundane and transcendental courses of action and thinks of pragmatic tools. He once gave a simile of water and raft to get an appropriate solution for any impending issue. During a *saṁsāric* long journey, a person should choose his action wisely as in a great expanse of water with a frightening and dangerous shore, a wise person collects the grass, and a piece of wood, binding them together to make a raft. He crossed the river with the help of the raft.[73] Masaru Emoto says that the more you learn about the nature and characteristics of water, the more difficult it will be for a person to deny the existence of the divine. The person has to realise that one's birth and existence are due to water. In the embryonic stage, the human body consists of 90 percent water; after birth, it comes to 70 percent in adulthood. When humans age, the percentage of water recedes in the body to 50 percent. So, in living existence, humans are nothing but water, and like human life, the nature of water is momentary. When organs run, the body is fit to respond, and when organs stop working, it leads to decay and death. Similarly, when water is running, it is considered pure, and in case of stagnant, it develops deformities.[74] Masaru Emoto examines water's physical and social characteristics through the formation of water crystals. He proposes that the crystal becomes distorted and impure when water is polluted or defiled. Still, in pure water, the structure of the crystal becomes clearer with some noble characteristics. Like other living beings, when one is suffering from an issue, the radiated energy and vibes will be negative and low, and in the case of happiness, the outcome will be positive and vibrant. Plants, trees, and water also have senses and respond accordingly in good or bad times. Some people can listen and understand these senses.[75] Water has a unique ability to absorb and dissolve other things and carry them away. It is linked to the regenerative powers of the water. Still, in contemporary times, millions of tons of industrial wastes, chemicals, and human excreta are thrown in the water, and water takes a long period to regenerate itself. Some of the materials are even indissoluble and pollute the water. The ability of water to absorb other substances is equated with the notion of the 'soup of life' that supplies the lakes, rivers, and oceans with the necessary nutrients to sustain the life therein. Because of these characteristics, water is considered the source of genesis and creation in many ancient civilisations.[76] The moon has a substantial effect on the physical nature of water and its propensity can be seen in the movements of oceanic waves. Joan S. Davis says that the moon affects not only the external water but also water and liquid contents in living beings. Since ancient times, the cyclical changes in seasons and climate have been a cause of inquiry. Meteorological studies reveal water's attributes affecting the periodicities seen in abiotic and biotic systems. The studies also show that planetary movements and fluctuations also affect our health patterns, and the role of water here is to serve as a vehicle of communication to both biotic and abiotic systems. The relationship between the abiotic system, like the moon, another planetary system, and humans show that living beings constantly face changes due to the changing nature of gravitational and other physical forces. The inherent climatic changes due to this continuously changing moon cycle give different rain patterns and economic cycles. The people of ancient civilisations were wise enough to learn different moon phases and zodiac constellations and develop a set pattern of agriculture based on this information. Because of that, they believed in the sacred nature of water and water as the bestower of life. Science has been so disenchanted to discover and admit the lunar effects upon the Earth and its inhabitants because of an inadequate understanding of water. The contemporary research on water structure and the pattern of its behaviour will strengthen the belief that water can communicate information between the Moon and Earth. It will also precede acknowledging the contribution of the subtle, life-supporting mechanism the Moon delivers to our planet.[77] The miscellaneous marvels of planet Earth have changed since its evolution, and the contemporary geomorphic landscapes have appeared due to different intergalactic and natural incidences over millions of years. With the increasing population and ever-changing structure of the globalised world, the *Homo sapiens's* association with water has kept changing. Tambiah recounts that a procedure of rethinking in Buddhism has already happened where monks uphold their traditional monastic beliefs and engage in 'Protestant ethics, which existed in this worldly asceticism'.[78] The engaged tradition in Buddhism has given an innovative outlook to Buddhism and a generous interpretation of its welfare economics and governance teachings. The theory of dependent origination now determines that there is no accidental cause for corrosion and disequilibrium for environmental crises. The condition has become alarming in the last hundred years, and vital attention is required to introspect human actions linked to exploiting natural resources. It must be free from gluttony, greed, and misapprehension. The ancient civilisational instinct that water and other natural resources are sacred, regenerative, and social will be the ideal path to coexist with them most peacefully and productively.

Notes

1 Veronica Strang, *Water Nature and Culture*, London: Reaktion Books, 2015, pp. 26–27.
2 Anand Singh, *Planet, Plants, and Animals Ecological Paradigms in Buddhism*, New Delhi: Primus Books, 2019, pp. 229–230.
3 Anand Singh, 'Water Sources in the Buddhist Ecology: Looking through the Engaged Tradition', *Journal of Social Innovation and Knowledge*, vol. 1, 2024, pp. 65–87.
4 Terje Tvedt and Eva Jakobsson, 'Introduction: Water History is World History', in *A History of Water, Water Control and River Biographies*, Vol. 1, ed. Terje Tvedt and Eva Jacobson, London: I.B. Tauris, 2006, pp. ix–xii.
5 Singh, *Planet, Plants, and Animals*, p. 1.
6 Peter Timmerman, 'Western Buddhism and Global Crisis', in *Dharma Rain: Sources of Buddhist Environmentalism*, ed. Stephanie Kaza and Kenneth Kraft, Shambhala: Shambhala Publications, 2000, p. 357.
7 Bhikkhu. Bodhi, 'A Buddhist Response to Contemporary Dilemmas of Human Existence', 1994, http://www.accesstoinsight.org/lib/authors/bodhi/response.html.
8 Bridget Allchin, 'Early Man and Environment in South Asia 10000 BC–AD 5000', in *Nature and the Orient*, ed. Richard H. Grove, Vinita Damodaran, and Satpal Narayan, New Delhi: Oxford University Press, 1997, pp. 30–31.
9 Geroge Erdosy, 'Deforestation in Pre and Protohistoric South Asia', in *Nature and the Orient*, ed. Richard H. Grove, Vinita Damodaran, and Satpal Narayan, New Delhi: Oxford University Press, 1997, pp. 64–65.
10 G.P. Marsh, *Man and Nature*, New York: Scribner Press, 1965, p. 9.
11 Singh, *Plants, Planet and Animals*, pp. 6–7.
12 Andrew Goudie, *The Human Impact on the Natural Environment*, Oxford: Basic Blackwell Ltd., 1990, pp. 4–8.

13 White Lynn Jr. 'The Historical Roots of Our Ecological Crisis', *Science*, vol. 155, 1967, p. 1206.
14 James M. Gustafson Gustafson, *A Sense of Divine: The Natural Environment From a Theocentric Perspective*, Cleveland: Pilgrim Press, 1994, pp. 78–90.
15 Donald K. Swearer, 'An Assessment of Buddhist Eco Philosophy', *Harvard Theological Review*, vol. 99, no. 2, 2006, pp. 123–124.
16 Cristopher S. Queen, 'Introduction: The Shapes and Sources of Engaged Buddhism', in *Engaged Buddhism; Buddhist Liberation Movements in Asia*, ed. Christopher S. Queen and Sallie B. King, Albany: State University of New York, 1996, pp. 23, 30.
17 James E. Deitrick, 'Engaged Buddhist Ethics: Mistaking the Boat for Shore', in *Action Dharma: New Studies in Engaged Buddhism*, ed. Christopher S. Queen, Charles Prebish and Damien Keown, London: Routledge, 2003, pp. 252–269.
18 S. J. Tambiah, 2013. 'The Galactic Polity in Southeast Asia', *HAU: Journal of Ethnographic Theory*, vol. 3, no. 3, 2013, p. 503.
19 Juliane Schober, 'The Theravāda Buddhist Engagement with Modernity in Southeast Asia: Whither the Social Paradigm of the Galactic Polity?', *Journal of Southeast Asian Studies*, vol. 26, no. 2, 1995, p. 308.
20 S. J. Tambiah, *World Conqueror and World Renouncer, A Study of Buddhism and Polity in Thailand Against a Historical Background*, Cambridge: Cambridge University Press, 1976, pp. 422–423.
21 Singh, *Planet, Plants, and Animals*, pp. 24–25.
22 England Philippa, 'UNCED and the Implementation of Forest Policy in Thailand', in *Seeing Forest for Tress: Environment and Environmentalism in Thailand*, ed. Philip Hirsch, Chaing Mai: Silkworm Books, 1996, pp. 53–71.
23 Pipob Udomittipong, 'Thailand's Ecology Monks', in *Dharma Rain: Sources of Buddhist Environmentalism*, ed. Stephanie Kaza and Kenneth Kraft, Boston: Shambala, 2000, p. 191.
24 Udomittipong, 'Thailand's Ecology Monks', pp. 197–199.
25 Susan M. Darlington, 'The Ordination of a Tree: The Buddhist Ecology Movement in Thailand', *Ethnology*, vol. 37, no. 1, 1998, pp. 1–15.
26 Seth Devere Clippard, 'The Lorax Wears Saffron: Towards a Buddhist Environmentalism', *Journal of Buddhist Ethics*. 2011, pp. 228–229.
27 Darlington, 'The Ordination of a Tree', pp. 7–10.
28 Susan M. Darlington, 'Not Only Preaching- The Work of the Ecology Monk Phrakhru Pitak Nanthakhun of Thailand', *Forest Trees and People, Newsletter*, vol. 34, no. September, 1997, pp. 17–20.
29 Singh, *Planet Plants and Animals*, p. 25.
30 Allchin, 'Early Man and Environment in South Asia 10000 BC–AD 500', pp. 31–33.
31 Gerorge Erdosy, 'Deforestation in Pre and Protohistoric South Asia', in *Nature and the Orient*, ed. Richard H. Grove, Vinita Damodaran, and Satpal Sangwan, New Delhi: Oxford University Press, 1998, pp. 56–57.
32 Dinesh Pathak and Ananta P. Gajurel, *Climate Change Impacts on Hazards in the Eastern Himalayas*, Kathmandu: International Centre for Integrated Mountain Development, 2010, pp. 5–7.
33 Anand Singh, 'Ecological and Environmental Concerns Amongst the Buddhist Communities of Ladakh: A Case Study', in *Buddhist Virtues and Socio-Economic Development*, MCU Edited Volume, Bangkok, 2011, p. 432.
34 Davide Geneletti and Dawa, Dorje, 'Environmental Impact Assessment of Mountain Tourism in Developing Regions: A Study in Ladakh, Indian Himalaya', *Environmental Impact Assessment Review*, vol. 29, 2009, pp. 230–233.
35 Tsetan Dolker, 'Impact of Ice Stupa in Mitigating the Obstacles to Agriculture in Ladakh', *Journal of Emerging Technologies and Innovative Research, JETIR*,

vol. 10, no. 4, 2023, pp. 366–395; Chewang Norphel and Padma Tashi,. 'Snow Water Harvesting in the Cold Desert in Ladakh: An Introduction to Artificial Glacier', in *Mountain Hazards and Disaster Risk Reduction. Disaster Risk Reduction*, ed. H. Nibanupudi, and R. Shaw, Tokyo: Springer, 2015, pp. 199–210 https://doi.org/10.1007/978-4-431-55242-0_11.

36 Wangchuck Sonam, 'Ice Stupa Artificial Glaciers of Ladakh', 2014, https://www.indiegogo.com/projects/ice-stupa-artificial-glaciers-of-ladakh; C. Clouse, 'Frozen Landscapes: Climate-Adaptive Design Interventions in Ladakh and Zanskar', *Landscape Research*, vol. 41, no. 8, 2016, pp. 821–837.

37 Shraddha Kumar, A *Self-Governance Approach to Solving the Water Crisis in Ladakh, India: The Ice Stupa Project*, Enschede, Netherlands: University of Twente (Unpublished Master Thesis), 2018–2019, pp. 7–8.

38 C. Clouse, 'The Himalayan Ice Stupa: Ladakh's Climate Adaptive Water Cache', *Journal of Architectural Education*, vol. 71, no. 2, 2017, pp. 248–249. https://doi.org/10.1080/10464883.2017.1340781; C. Clouse, N. Anderson and T. Shippling,'Ladakh's Artificial Glaciers: A Climate-adaptive Design for Water Scarcity', *Climate and Development*, vol. 9, no. 5, 2017, pp. 428–438. https://doi.org/10.1080/17565529.2016.1167664.

39 Janaki Lenin, "Artificial Glacier Could Help Ladakh Villagers Adapt to Climate Change', *The Guardian*, 2015. https://www.theguardian.com/environment/2015/feb/24/artificial-glacier-could-help-ladakh-villagers-adapt-to-climate-change.

40 Peter Aldhous, 'The World's Forgotten Crisis', *Nature*, vol. 422, 2003, p. 251. https://doi.org/10.1038/422251a.

41 Masaru Emoto, *The Hidden Messages in Water*, Oregon: Beyond Words Publishing, 2001, pp. 62 63; Davis, 'Water: Communicator in Moon-Earth Relationships', 2001, p. 127.

42 Jamie Linton, *What Is Water? The History of a Modern Abstraction*, Chicago: The University of Chicago Press, 2010, pp. 51–52.

43 *The Report of the World Commission on Dams: Dams and Developments- Dams and Development, A New Framework*, London and Sterling: Earthscan Publications Ltd, 2000, p. 8.

44 D. Worster, *Rivers of Empire: Water, Aridity, and the Growth of American West*, New York: Pantheon, 1985, p. 5.

45 K. DiFrancesco, 'The Debate over Large Dams', in *Global Perspectives on Large Dams: Evaluating the State of Large Dam Construction and Decommissioning across the World: Report on a Conference held on 3–5 November 2006 at the Yale School of Fraternity and Environmental Studies, Yale School of Fraternity and Environmental Studies Report*, ed. K. DiFrancesco and K. Woodruff, New Haven CT: Yale School of Forestry and Environmental Science, 2008, p.7.

46 P.H. Gleick, 'A Look at Twenty-First Century Water Resources Development', *Water International*, 2000, vol. 25, no. 1, pp. 127–138. DOI 10.1080/02508060008686804.

47 Mathew D. Evenden, *Fish versus Power: An Environmental History of the Fraser River*, Cambridge: Cambridge University Press, 2004, p. 1.

48 Aldhous, 'The World's Forgotten Crisis', p. 251.

49 Linton, *What is Water?*, pp. 193–194.

50 Ken Conca, 2006. *Governing Water Contentious Transnational Politics and Global Institution Building*, Cambridge MA: MIT Press, 2006, pp. 1–2.

51 Linton, *What Is Water?*, pp. 6–7, 18.

52 Alfred Bloom. *Globalization and Buddhism*, 2012, http://www.shindharmanet.com/wpcontent/uploads/2012/pdf/Bloom-Globalization.pdf.

53 Jan Tinbergen, 'Wise Management for More Human World', in *Solutions for a Troubled World*, ed. Marc Macy, London: Earth view Press, 1987, pp. 56–57.

54 E.F. Schumacher, 'Buddhist Economics', *Manas*, vol. XXII, no. 33, 1969, pp. 1–14. www.manasjournal.org/pdf.; Ananda W.P. Gurge, 'Buddhist Economics-Myth and Reality', *Hsi Lai Journal of Humanistic Buddhism*, vol. 7, 2006, pp. 116–117. http://journal.uwest.
55 Stephanie Kaza and Kenneth Kraft, *Dharma Rain: Source of Buddhist Environmentalism*, Berkely: Shambhala Publication, 2000, p. xvii.
56 Ignacy Sachs, 'Development, Maldevelopment and Industrialization of Third World Countries', *Development and Change*, vol. X, no. 4, 1979, pp. 635–46.
57 Anthony Le Duc, 'Theravada Buddhist Environmental Spirituality: Relational and Developmental Dimensions in Promoting Environmental Well-Being', *Verbum SVD*, vol. 57, no. 3–4, 2016, p. 366.
58 *Digha Nikāya*, III. 58.
59 Frank Reynolds, "The Two Wheels of Dhamma: A Study of Early Buddhism," in *The Two Wheels of Dhamma: Essays on the Theravada Tradition in India and Ceylon*, ed. Gananath Obeyesekere, Frank Reynolds and Bardwell L. Smith, A.A.R. Studies in Religion Series, no. 3 Chambersberg: A.A.R. 1972, p. 13.
60 Tambiah, *World Conqueror and World Renouncer*, pp. 39–40.
61 *Majjhima Nikāya*, II.179.
62 *Vinaya Pitaka*, I.10; *Saṁyutta Niakāya*, V.421.
63 David R. Loy, *The Great Awakening: A Buddhist Social Theory*, Boston: Wisdom Publications, 2003, pp. 1–3.
64 Loy, *The Great Awakening*, pp. 19–28.
65 *Saṁyutta Nikāya*, II.25.
66 Joanna R Macy, 'Dependent Co-arising: The Distinctiveness of Buddhist Ethics', *The Journal of Religious Ethics*, Spring, vol. 7, no. 1, 1979, pp. 39–40.
67 Frank, Reynolds. 'Multiple Cosmogonies and Ethical Order: The Case of Theravada Buddhism', in *Cosmogony and Ethical Order*, ed. Robin W. Lovin and Frank Reynolds, Chicago: University of Chicago Press, 1985, p. 207; Reynolds Frank, 'Four Modes of Theravada Action', *Journal of Religious Ethics*, vol. 7, no. Spring, 1979, pp. 12–26.
68 Reynolds, 'Multiple Cosmogonies and Ethical Order', p. 208.
69 Bhikkhu Bodhi, 'A Buddhist Response to Contemporary Dilemmas of Human Existence', 1994, http://www.accesstoinsight.org/lib/authors/bodhi/response.html.
70 David J. Kalupahana, *Buddhist Philosophy: A Historical Analysis*, Hawaii: The University Press of Hawaii, 1976, p. 30.
71 Payutto, 'Buddhist Economics', pp. 1–3.
72 Simon Zadek, 'The Practice of Buddhist Economics? Another View', *American Journal of Economics and Sociology*, vol. 52, no. 4, 1993, p. 433, http://www.jstor.org/stable/3487468.
73 *Majjhima Nikāya*, I.134–135.
74 Masaru Emoto, *The Hidden Messages in Water*, New York: Atria Books, 2005, pp. x–xv.
75 Emoto, *The Hidden Messages in Water*, pp. 40–43.
76 Emoto, *The Hidden Messages in Water*, pp. 55–56.
77 Joan S. Davis, 'Water: Communicator in Moon-Earth Relationships', in *Earth-Moon Relationships*, ed. C. Barbieri and F. Rampazzi, Springer: Dordrecht, 2001, p. 127, https://doi.org/10.1007/978-94-010-0800-6_12.
78 Tambiah, *World Conqueror and World Renouncer*, pp. 427–428.

Bibliography

Primary Sources

Abhiniṣkramanasūtra, tr. Samuel Beal. *The Romantic Legend of Śākya Buddha, A Translation of Chinese Version of the Abhiniṣkramanasūtra*, New Delhi: Samyak Prakashan, 2011, reprint.

Acarānga Sūtra, tr. Herman Jacobi, *Jain Sutras*, 2 vols., New Delhi: Motilal Banarasidas, 1968, reprint.

Aitareya Brahmana of the *Ṛigveda*, tr. Martin Haug. Allahabad: Sudhindra Nath Vasu & Co, 1922.

Aṅguttara Nikāya. tr. Bhikkhu Bodhi, *The Numerical Discourses of the Buddha: A Complete Translation of the Aṅguttara Nikāya (The Teachings of the Buddha)*, Shambhala: Wisdom Publication, 2012.

Aṅguttara Nikāya. The Bood of Gradual Saying, tr. F.L. Woodward, New Delhi: Motilal Banarasidas, 2006.

Apadāna, Apadanapali. Legend of the Buddhist Saints, tr. Jonathan S. Walter, 2018, apadanatranslation.org

Arthaśāstra, of Kautilya, ed. R. Shamasastri, Mysore: Wesleyan Mission Press, 1919.

Aṭṭhaśālinī of Buddhaghosha, tr. E.I. Muller, London: Pali Text Society, 1897.

Beal, Samuel, *Life of Hieun-Tsiang by the Shaman Hwui Li*, London: K. Paul, Trench & Trubner, 1911.

Beal, Samuel, *Travels of Fah-Hian and Sung-Yun from China to India* (*400 AD & 518 AD*), Delhi: Low Price Publication, 2005, reprint.

Buddha-carita, or Life of Buddha Venerable Aśvaghoṣa, tr. E.B. Cowell, Oxford: The Clarendon Press, 1893.

Chaṅdogya Upaniṣada, tr. Ganganath Jha, Poona: Oriental Book Agency, 1942.

Chavannes, Edouard, *Memoire Campose a lepage de la Grande Dyanasties Tang sur les Religieux Eminents par I-tsing* (*Kan-fa-kuosang-chuan*), Paris: E Leroux, 1894.

Das, Sarat Chandra, ed., *Pag Sam Jon Zang by po yece pal jor as History of Rise, Progress, and Downfall of Buddhism in India*, Calcutta: Princeton Jail Press, 1908.

Dhammapadaṭṭhakathā, ed. C.S. Upasaka, Nalanda: Nav Nalanda Mahavihara, 1973.

Dīgha Nikaya, tr. Maurice Walshe, *The Long Discourses of the Buddha: A Translation of the Digha Nikaya* (The Teachings of the Buddha), Boston: Wisdom Publication, 1995.

Dīgha Nikāya. Dialogues of the Buddha: English Translation of Digha Nikaya, tr. T.W. Rhys Davids and C.A.F. Rhys Davids, 3 Vols. New Delhi: Motilal Banarasidas, 2010, reprint.

Divyavadāna, or A Collection of Early Buddhist Legends, ed. Edward B. Cowell and Robert A. Neil, Amsterdam: Oriental Press, 1970.

Giles, H.A. *Record of the Buddhistic Kingdoms: Translated from Chinese*, London: Trubner and Co., 1918.

Harsacarita, Bombay, tr. E.B. Cowell and F.W. Thomas, London: Royal Asiatic Society, 1897.
Jātaka, or the Stories of the Buddha's former Birth, tr. E.B. Cowell, New Delhi: Low Price Publication, 1993, reprint.
Kalpasūtra, tr. J. Stevenson, *The Kalpa Sutra, and Nava Tatva: Two Works Illustrative of the Jain Religion and Philosophy*, New Delhi: Forgotten Books, 2018.
Lalitavistara, tr. Rajendralal Mitra, Calcutta: Asiatic Society, 2010, reprint.
Lalitavistara, ed. S. Lefmann, *Lalita Vistara, Leben und Lehre des Śakya-Buddha, Textausgabe mit Varianten*, Halle: Metrenund Wörterverzeichnis, 1902.
Lama, Chimpa and A. Chattopadhyaya, *Taranatha's History of Buddhism in India*, Shimla: Indian Institute of Advanced Studies, 1970.
Legge, J., tr., *The Travels of Fa-hien (399–414 AD) or Record of the Buddhist Kingdoms*, Oxford: Clarendon Press, 1886.
Mahābhārata, tr. K.M. Ganguli, *The Mahabharata of Krishna-Dwaipayana Vyasa*, translated into English Prose from the Original Sanskrit Text, 11 vols., New Delhi: Munshiram Manoharlal Publishers, 1990, reprint.
Mahāvaṁsa, Great Chronicle of Ceylon, tr. Wilhelm Geiger, Delhi: Asian Educational Service, 1996, reprint
Mahāvastu, tr. J.J. Jones, London: Luzac and Company, vol. 1, 1949.
Maitrāyanī Saṁhitā, tr. Leopold Von Schroeder, vol. 1, Leipzig: F.A. Brockhaus, 1885.
Majjhima Nikāya. The Collection of Middle Length Saying, tr. I.B. Horner, New Delhi: Motilal Banarasi Das, 2004.
Manorathapūranī of Buddhaghosa, ed. Max Walleser, London: Pali Text Society, 1924.
Matsya Purāṇa, tr. A. Taluqdar of Oudh, New Delhi: Munshiram Manoharlal, 1980.
Megha Sūtra, tr. Cecil Bendall. 'The Megha Sutra', *The Journal of the Royal Society of Great Britain and Ireland*, New Series, vol. 12, no. 2, 1880, pp. 286–311, www.jstore.org/stable/25196850/
Milindapañha, The Question of King Milinda, tr. T.W. Rhys Davids, vol. 1, New Delhi: Low Price Publication, 2006, reprint.
Papañcasudānī of Buddhaghosha, ed. J.H. Wood and D. Kosambi, London: Pali Text Society, 1922.
Rāmāyana, of Vālmṛki, tr. Ralph T.H. Griffith, London: Trubner and Co., 1886.
Ṛigveda, tr. T.H. Griffith, *The Hymns of the Rigveda*, 1896, http://www.sanskritweb.net/rigveda/griffith-p.pdf
Ṛigveda, Samhita and Pada Texts with Sayana's Commentary, tr. F. Max Muller, New Delhi: Low Price Publication, 2006, reprint.
Roerich, George, *Biography of Dhammasvami (Chag lo tsa-ba chos rje-dpal): A Tibetan Monk Pilgrim*, Patna: K.P. Jaiswal Research Institute, 2016, reprint.
Samantapāsādikā, ed. J. Takakusu and M. Nagai, vol. 1, London: Pali Text Society, 1924–47.
Saṁyutta Nikāya, tr. Bhikkhu Bodhi, *The Connected Discourse of the Buddha: A Translation of the Samyutta Nikaya*, Boston: Wisdom Publication, 2003.
Saṁyutta Nikāya. The Book of Kindred Sayings, tr. T.W. Rhys Davis and F.L. Woodward. New Delhi: Motilal Banarasidas, 2005. Reprint.
Saratthapakāsinī of Buddhaghosa, ed. F.L. Woodward, vol. 1, London: Pali Text Society, 1977.
Śatapatha Brahmana, Sanskrit Text with English Translation, Notes, Introduction and Index of Julius Eggeling, 3 Volume, New Delhi: Eastern Book Linkers, 2009, reprint.
Suttanipāta Commentary, ed. Helmer Smith, London: Pali Text Society, 1917.
Takakusu, J., tr. I. Tsingh, *A Record of the Buddhist Religion as Practiced in India & Malay Archipelago (AD 671–695)*, New Delhi: Munsiram Manoharlal, 1998.
Theragatha Atthakatha, or *Paramattha-dipani Theragatha-atthakatha: the Commentary of Dhammapala-cariya*, ed. F.L. Woodward, London: Pali Text Society, 1971.

Theragatha. Psalms of the Brethren, tr. C.A.F. Rhys Davids, Oxford: Pali Text Society, 1913.

Therigatha. Verses of Elder Nuns, tr. C.A.F. Rhys Davids and K.R. Norman, Oxford: Pali Text Society, 1989.

Udana Commentary, Paramatthadipani nama Udanatthakatha of Dhammapala, tr. Peter Masefield, 2 vols., Oxford: Pali Text Society, 1994–95.

Udāna or *Verses of Uplift* in *Minor Anthologies*, Vol. II, tr. P. Steinthal, London: Pali Text Society, 2017.

Vimanavatthu Atthakatha, tr. Peter Masefield and N.A. Jayawickrama, *Elucidation of the Intrinsic Meaning so named the Commentary on the Vimana Stories (Paramatthadipani nama Vimanavatthu-atthakatha)*, Oxford: Pali Text Society, 1989.

Vinaya Pitaka. The Book of Disciplines, tr. I.B. Horner. 6 Vols. London: Pali Text Society, 1949.

Vinaya-Vinicchaya, ed. D. Dhammarakshita, Colombo: Buddhist Publication Society, 1937.

Viṣnu Purāṅa, tr. K.L. Joshi, *The Visnu Purana: Sanskrit Text and English Translation of the Vishnu Purana*, New Delhi: Parimal Books, 2020.

Viśuddhimagga of Buddhaghosha, tr. Bhikshu Naṅamoli, *The Path of Purification*, Colombo: Buddhist Publication Society, 2010.

Vṛhatsaṁhitā, of Varāhamihira, tr. M.R. Bhatt, New Delhi: Motilal Banarsidass, 1987.

Watters, Thomas, *On Yuan Chwang's Travels in India (AD 629–645)*, 2 vols., New Delhi: Low Price Publication, 2004, reprint.

Secondary Sources

Aboshi, Y., K. Sonada, F. Yoneda, and A. Uesugi. 'Excavation at Sahet Mahet 1986–1996', *East and West*, vol. 49, no. 1/4, 1999, pp. 119–173.

Aldhous, Peter. 'The World's Forgotten Crisis', *Nature*, vol. 422, 2003, p. 251. https://doi.org/10.1038/422251a

Allchin, Bridget. 'Early Man and Environment in South Asia 10000 BC–AD 500', in *Nature and the Orient*, ed. Richard H. Grove, Vinita Damodaran, and Satpal Narayan, New Delhi: Oxford University Press, 1997, pp. 29–59.

Andrew, Goudie. *The Human Impact on the Natural Environment*, Oxford: Basic Blackwell Ltd, pp. 1990.

Ann, Heirman and Mathieu Torck, *A Pure Mind in a Clean Body, Body Care in Buddhist Monasteries of Ancient Indian and China*, Gent: Academia Press, 2012.

Balasubramanian, A. 'The World's Water', 2015. https://www.researchgate.net/publication/315123891

Barth, Auguste. *The Religions of India*, Boston: Houghton Mifflin and Company, 1882.

Barua, Benimadh . *A History of Pre-Buddhistic Indian Philosophy*, Delhi: Motilal Banarasidas, 2016, reprint.

Bennett, John W. 'Anthropological Contributions to the Cultural Ecology and Management of Water Resources', in *Man and Water: The Social Sciences in Management of Water Resources*, ed. L. Douglas James, Lexington: University Press of Kentucky, 1974, pp. 34–81.

Bisht, R.S. *Excavation at Dholavira (1989–90 to 2004–05)*, 2015, pp. 921–922, asi.nic.in/asi_exca_2007_dholavira.asp

Biswas, Asit K. *History of Hydrology*, Amsterdam: North Holland Publishing, 1970.

Blackburn, Anne M. 'Magic in the Monastery: Textual Practice and Monastic Identity in Sri Lanka', *History of Religions*, vol. 38, no. 4, 1999, pp. 354–372.

Bloom, Alfred. 'Globalization and Buddhism', 2012, http://www.shindharmanet.com/wpcontent/uploads/2012/pdf/Bloom-Globalization.pdf

Bloss, Lowell W. 'The Buddha and the Naga: A Study in Buddhist Folk Religiosity', *History of Religions*, vol. 13, no. 1, 1973, pp. 36–53.

Bodhi Bhikkhu. 'A Buddhist Response to Contemporary Dilemmas of Human Existence', 1994, http://www.accesstoinsight.org/lib/authors/bodhi/response.html

Boton, Carol Radcliffe. 'Images of Agastya in Nepal', *Artibus Asiae*, vol. 51, no. 1/2, 1991, pp. 75–89.

Boucher, Daniel. 'Sutra on the Merit of Bathing the Buddha', in *Buddhism in Practice*, ed. Donald S. Lopez Jr., Princeton: Princeton University Press, 1995, pp. 59–68.

Buhler, George. 'A New Kharoshthi Inscription from Swat', *The Indian Antiquary*, vol. XXV, 1896, pp. 141–142.

Chattaraj, Arpita. 'Sacred Water and Cultures of Worship: Some Observations on the River in India', *Humanities Bulletin*, vol. 4, no. 2, 2021, pp. 168–183.

Ciurtin, E. 'The Buddha's Earthquakes, On Water: Earthquakes and Seaquakes in Buddhist Cosmology and Meditation, with an Appendix on Buddhist Art', *Stvdia Asiatica*, vol. X, 2009, pp. 59–123.

Clippard, Seth Devere. 'The Lorax Wears Saffron: Towards a Buddhist Environmentalism', *Journal of Buddhist Ethics*, vol. 18, 2011, pp. 212–248.

Clouse, C. 'Frozen Landscapes: Climate-Adaptive Design Interventions in Ladakh and Zanskar', *Landscape Research*, vol. 41, no. 8, 2016, pp. 821–837.

Clouse, C. 'The Himalayan Ice Stupa: Ladakh's Climate Adaptive Water Cache', *Journal of Architectural Education*, vol. 71 no. 2, 2017, pp. 247–251, https://doi.org/10.1080/10464883.2017.1340781

Clouse, C., N. Anderson, and T. Shippling. 'Ladakh's Artificial Glaciers: A Climate-Adaptive Design for Water Scarcity', *Climate and Development*, vol. 9 no. 5, 2017, pp. 428–438, https://doi.org/10.1080/17565529.2016.1167664

Cohen, Richard S. 'Nagas Yaksini, Buddha: Local Deities and Local Buddhism at Ajanta', *History of Religions*, vol. 37, no. 4, 1998, pp. 360–400.

Colgrave, B. and R.A.B. Mynors. *Bede's Ecclesiastical History of the English People*, Oxford. University Press, 1969.

Conca, Ken. *Governing Water Contentious Transnational Politics and Global Institution Building*, Cambridge MA: MIT Press. 2006.

Coomaraswamy, Ananda K. *Yaksas*, New Delhi: Munshiram Manoharlal Publishers, 2001, reprint.

Cunningham, Alexander. *Four Reports Made During the Years 1862–63–64–65*, New Delhi: Archaeological Survey of India, 2000a.

Cunningham, Alexander. *Report of Tours in the Gangetic Provinces From Badaon to Bihar in 1875–76 and 1877–78*, New Delhi: Archaeological Survey of India, 2000b, reprint.

Cunningham, Alexander. *The Ancient Geography of India*, New Delhi: Munshiram Manoharlal, 2016 (reprint).

Darlington, Susan M. 'Not Only Preaching- The Work of the Ecology Monk Phrakhru Pitak Nanthakhun of Thailand', *Forest Trees and People*, *Newsletter*, vol. 34, no. September, 1997, pp. 17–20.

Darlington, Susan M. 'The Ordination of a Tree: The Buddhist Ecology Movement in Thailand', *Ethnology*, vol. 37, no. 1, 1998, pp. 1–15.

Davis, Joan S. 'Water: Communicator in Moon-Earth Relationships', in *Earth-Moon Relationships*, ed. C. Barbieri and F. Rampazzi, Dordrecht: Springer, 2001. https://doi.org/10.1007/978-94-010-0800-6_12

de Silva, Lily. *Paritta: A Historical and Religious Study of the Buddhist Ceremony for Peace and Prosperity in Sri Lanka*, Colombo: National Museum, 1981.

de Viser, M.W. *The Dragon in China and Japan*, Amsterdam: Johannes Muller, 1913.

Decaroli, Robert. *Haunting the Buddha*, New York: Oxford University Press, 2004.
Decaroli, Robert. 'Snakes and Gutters Naga Imagery, Water Management, and Buddhist Rainmaking Rituals in Early South Asia', *Archives of Asian Art*, vol. 69, no. 1, 2019, pp. 1–19.
Deeg, Max. *Miscellanae Nepalicae Early Chinese Reports on Nepal, the Foundation Legend of Nepal in its Trans-Himalayan Context*, Lumbini: Lumbini International Research Institute, 2016.
Deitrick James E. 'Engaged Buddhist Ethics: Mistaking the Boat for Shore', in *Action Dharma: New Studies in Engaged Buddhism*, ed. Christopher S. Queen, Charles Prebish and Damien Keown, London: Routledge, 2003, pp. 252–269.
Deshpande, M.N. 'The Rock Cut Caves of Pitalkhora in the Deccan', *Ancient India*, vol. 15, 1959, pp. 66–93.
DiFrancesco, K. 'The Debate over Large Dams', in *Global Perspectives on Large Dams: Evaluating the State of Large Dam Construction and Decommissioning across the World: Report on a Conference held on 3–5 November 2006 at the Yale School of Fraternity and Environmental Studies, Yale School of Fraternity and Environmental Studies Report*, ed. K. DiFrancesco and K. Woodruff, New Haven, CT: Yale School of Forestry and Environmental Science, 2008, pp. 7–14.
Dolker, Tsetan. 'Impact of Ice Stupa in Mitigating the Obstacles to Agriculture in Ladakh', *Journal of Emerging Technologies and Innovative Research*, vol. 10, no. 4, 2023, pp. 366–395.
Donahue, J.M. and B.R. Johnston. 'Conclusion', in *Water, Culture, and Power: Local Struggle in Global* Context, ed. J.M. Donahue and B.R. Johnston, Washington, DC: Island Press, 1998, pp. 339–346.
Dowden, Ken. *European Paganism: The Realities of Cult from Antiquity to the Middle Ages*, New York: Routledge, 2002.
Dutt, Sukumar. *Buddhist Monks and Monasteries of India: Their History and Their Contribution to Indian Culture*, New Delhi: Motilal Banarasidas, 2010.
Eck, Diana L. 'Ganga: The Goddess Ganges in Hindu Sacred Geography', in *An Anthology of Writings on the Ganga: Goddess and River in History, Culture, and Society*, ed. Assa Doron, Richard Barz, and Barbara Nelson, New Delhi: Oxford University Press, 2015, pp. 233–251.
Eliade, Mircea. *Patterns in Comparative Religion*, London and New York: Sheed and Ward, 1958.
Emmirick, R.E. *The Sutra of Golden Light, Being a Translation of the Suvarnaprabhasa Sutra*, Oxford: Pali Text Society, 2004.
Emoto, Masaru. *The Hidden Messages in Water*, Oregon: Beyond Words Publishing, 2001.
England, Philippa. 'UNCED and the implementation of Forest Policy in Thailand', in *Seeing Forest for Tress: Environment and Environmentalism in Thailand*, ed. Philip Hirsch, Chaing Mai: Silkworm Books, 1996, pp. 53–71.
Erdosy, Gerorge. 'Deforestation in Pre and Protohistoric South Asia', in *Nature and the Orient*, ed. Richard H. Grove, Vinita Damodaran, and Satpal Sangwan, New Delhi: Oxford University Press, 1998, pp. 51–69.
Evenden, Mathew D. *Fish versus Power: An Environmental History of the Fraser River*, Cambridge: Cambridge University Press, 2004.
Farrington, B. *Greek Science*, Harmondsworth: Penguin Books, 1963.
Fiordalis, David V. 'Miracles in Indian Buddhist Narratives and Doctrine', *Journal of the International Association of Buddhist Studies*, vol. 33, no. 1–2, 2010, pp. 381–408.
Frude, Laura D. *Verbal Magic in Pali Canon, A Study on Protective Chants, Truth-acts, and Vows*, Bristol: University of Bristol, 1991, pp. 21–22, http://research-informationbristol.ac.uk

Gargan, Edward. *The River's Tale: A Year on Mekong*, New York: Vintage Books, 2002.

Geneletti, Davide and Dorje Dawa. 'Environmental Impact Assessment of Mountain Tourism in Developing Regions: A Study in Ladakh, Indian Himalaya', *Environmental Impact Assessment Review*, vol. 29, 2009, pp. 229–242.

Gethin, R.M.L. *Sayings of the Buddha: A Selection of Suttas from Pali Nikāyas*, Oxford: Oxford University Press, 2008.

Gleick, P.H. 'A Look at Twenty-First Century Water Resources Development', *Water International*, vol. 25, no. 1, 2000, pp. 127–138. https://doi.org/10.1080/02508060008686804

Gokhale, S. *Kanheri Inscriptions*, Pune: Deccan College, 1991.

Gombrich, Richard. *Precept and Practice, Traditional Buddhism in the Rural Highlands of Ceylon*, Oxford: Clarendon Press, 1971.

Gombrich, Richard. 'The Buddhist Attitude to Thaumaturgy', in *Bauddhavidyāsudhākaraḥ. Studies in Honour of Heinz Bechert on the Occasion of His 65th Birthday*, ed. Petra Kieffer-Pülz, and Jens-Uwe Hartmann, Swisttal-Odendorf: Indica et Tibetica Verlag, 1997, pp. 165–185.

Gomperz, H. 'Problems and Methods of Early Greek Science', *Journal of the History of Ideas*, vol. 4, 1943, pp. 161–176.

Greene, Paul D. 'The Dhamma as Sonic Praxis: Paritta Chant in Burmese Theravada Buddhism', *Asian Music*, vol. 35, no. 2, 2004, pp. 43–78.

Griffith, Paul J. 'A Hymn in Praise to Buddha's Good Qualities', in *Buddhism in Practice*, ed. Donald S. Lopez Jr., Princeton: Princeton University Press, 2007, pp. 39–49.

Griffith, Paul J., Noriaki Hakamaya, John P. Keenan, and Paul L. Swanson. *The Realm of Awakening: A Translation and Study of the Tenth Chapter of Asnga's Mahayanasangraha*, New York: Oxford University Press, 1989, pp. 123–168.

Gunaratna, V.F. 'Foreward', in *The Book of Protection*, Piyadassi Thera, Kandy: Buddhist Publication Society, 1999, pp. 11–21.

Gurge, Ananda W.P. 'Buddhist Economics-Myth and Reality', *Hsi Lai Journal of Humanistic Buddhism*, vol. 7, 2006, pp. 72–130. http://journal.uwest

Gustafson, James M. *A Sense of Divine: The Natural Environment From a Theocentric Perspective*, Cleveland: Pilgrim Press, 1994.

Guthrie, W.K.C. *A History of Greek Philosophy, The Earlier Presocratic and the Pythagorean*, Cambridge: Cambridge University Press, vol. 1, 1979.

Gyaltsen, Khenpo. 'A Lamp Illuminating the Path to Liberation: An Explanation of Essential Topics for Dharma Students, Lhasey Lotsawa Translations', 2014, https://www.lhaseylotsawa.org

Hallisey, Charles. 'Auspicious Things', in *Buddhism in Practice*, ed. Donald S. Lopez Jr., Princeton: Princeton University Press, 1995, pp. 412–426.

Hardy, R. Spence. *A Manual of Buddhism, in its Modern Development*, London: Partridge and Oakey, 1853.

Hartel, H. *Excavation at Sonkh: 2500 Years of a Town in Mathura District*, Berlin: Dietrich Reimer Verlag, 1993, pp. 10–16; G.C. Chandra, 'Excavation at Rajgir', *ASIAR*, 1935–36, Delhi: ASI, 1938.

Harvey, Peter. 'The Dynamics of Paritta Chanting in Southern Buddhism', in *Love Divine: Studies in Bhakti and Devotional Mysticism*. ed. Karl Werner, Surrey: Curzon Press, 1983, pp. 53–84.

Harvey, Peter. 'Buddhist Reflections on "Consumer" and "Consumerism"', *Journal of Buddhist Ethics*, vol. 20, 2013, pp. 334–356.

Hashimi, N.H., R. Nigam, R.R. Nair and G. Rajagopalan. 'Holocene Sea-level Fluctuations on Western Indian Continental Margin: An Update', *Journal of Geological Society of India*, vol. 46, 1995, pp. 157– 162.

Hegewald, Julia A.B. 'The Lotus Pool: Buddhist Water Centuries in the Kathmandu Valley', *South Asian Studies*, vol. 13 no. 1, 1997, pp. 145–159.

Hidas, Gergely. *Mahāpratisaramahavidyārajnī, The Great Amulet, Great Queen of Spells, Introduction, Critical Edition, and Annotated Translation*, New Delhi: International Academy of Indian Culture and Aditya Prakashan, 2012.

Hidas, Gergely. *A Buddhist Ritual Manual on Agriculture Vajratuṇḍasamayakalparāja*, Critical Edition and Translation, Berlin/Boston: Walter de Gruyter, 2019.

Hoey, W. 'The Five Rivers of the Buddhists', *The Journal of Royal Asiatic Society of Great Britain and Ireland*, vol. 39, no. 1, 1907, pp. 41–47.

Hongsuwan, Pathon. 'Sacralization of the Mekong River through Folk Narratives', *Manushya: Journal of Humanities*, special issue, no. 19, 2011, pp. 33–45.

Ignacy, Sachs. 'Development, Maldevelopment and Industrialization of Third World Countries', *Development and Change*, vol. X, no. 4, 1979, pp. 635–646.

Inagaki, Hsao. *The Three Pure Land Sutras, The larger Sutra on Amitayus, The Sutra on Contemplation of Amitayus; The Small Sutra on Amitayus*, Berkely: Numata Centre for Buddhist Translation and Research, 1995.

Jacobs, Jeffrey W., Lewis Owen and Gilbert F. White. 'Mekong River', *Encyclopedia Britannica*, https://www.britannica.com/place/Mekong-River. Accessed 4 April 2023.

Jamie, Linton. *What Is Water? The History of a Modern Abstraction*, Chicago: The University of Chicago Press, 2010.

Jowett, B. *Dialogues of Plato*, London: Oxford University Press, vol. 2, 1892.

Kalupahana, David. *A History of Buddhist Philosophy: Continuities and Discontinuities*, Honolulu: University of Hawaii, 1992.

Kalupahana David J. *Buddhist Philosophy: A Historical Analysis*, Hawaii: The University Press of Hawaii, 1976.

Khatry, Prem K. 'Rain for the Drought: An Anthropological Inquiry into the Structure of the Buddhist Festival in Kathmandu Valley', *Contribution to Nepalese Studies*, vol. 23, no. 1, 1996, pp. 89–108.

Klein, Anne C. and Khetsun Sangpo. 'Hail Protection', in *Religions of Tibet*, ed. Donal S. Lopez Jr., Princeton: Princeton University Press, 1997.

Kulatilake, Cyril de Silva. 'Buddhist Chant in Sri Lanka and Its Musical Elements', *Jahrbuchfur Musikalishe Volksund Volkerkunde*, vol. 10, 1982, pp. 20–32.

Law, Bimla Churn. *Historical Geography of Ancient India*, New Delhi: Munshiram Manoharlal, 2016.

Le Duc, Anthony. 'Theravada Buddhist Environmental Spirituality: Relational and Developmental Dimensions in Promoting Environmental Well-Being', *Verbum SVD*, vol. 57, no. 3–4, 2016, pp. 336–370.

Lenin, Janaki. 'Artificial Glacier Could Help Ladakh Villagers Adapt to Climate Change,' *The Guardian*, 2015. https://www.theguardian.com/environment/2015/feb/24/artificial-glacier-could-help-ladakh-villagers-adapt-to-climate-change. Retrieved on 27 August 2023.

Lewis, Todd. 'A History of Buddhist Ritual', in *The Buddhist World*, ed. John Power, New York: Routledge, 2016, pp. 318–337.

Locke, S.J. and K. John. *Karunamaya: The Cult of Avalokieśvara Masyendranāth in the Valley of Nepal*, Kathmandu: Sahyogi Prakashan, 1980.

Lopez Jr., Donald S. *Buddhism in Practice*, Princeton: Princeton University, 1995.

Loy, David R. *The Great Awakening: A Buddhist Social Theory*, Boston: Wisdom Publications, 2003.

Ludvik, Catherine. *Recontextualizing the Praise of a Goddess: From Harivamsa to Yijing's Chinese Translation of the Sutras of Golden Light*, Kyoto: School of East Asian Studies, 2006.

Luis, Gomez. 'The Bodhisattva as Wonder-worker', in *Prajnaparamita and Related System: Studies in Honor of Edward Conze*, ed. Lewis R. Lancaster, Berkely: University of California, 1977, pp. 221–262.

Lynn Jr, White 'The Historical Roots of Our Ecological Crisis', *Science*, vol. 155, 1967, pp. 1203–1207.
Mackay, E.J.H. 'SD Area: The Great Bath and Adjacent Buildings', in *Mohenjo-daro and Indus Valley Civilization*, ed. John Marshall, London: Arthur Probasthain, 1931, vol. 3, pp. 131–150.
Macy, Joanna. 'Dependent Co-Arising: The Distinctiveness of Buddhist Ethics', *Journal of Religious Ethics*, 1979, vol. 7 no. 1, 1979, pp. 38–52.
Mahalingam, T.V. *The Nāgas in Indian History and Culture, Sir Subramanyam Ayyar Endowment Lecture, University of Madras*, Madras: G S Press, 1965.
Malalasekera, G.P. *Dictionary of Pali Proper Names*, 2 Vols. New Delhi: Motilal Banarasidas, 2007.
Mani, B.R. 'Excavation at Sankissa 1995–1996', *Puratattva*, 1995–96, no. 26, pp. 43–49.
Marsh, G.P. *Man and Nature*, New York: Scribner Press, 1965.
Marshall, John. 'The Buildings', in *Mohenjo-daro and Indus Valley Civilization*, ed. John Marshall, London: Arthur Probasthain, vol. 3, 1931, pp. 15–26.
Marshall, John. *The Monuments of Sanchi*, Calcutta: Survey of India, 1940.
Mate, M.S. 'Harappan Fortifications: A Study', in *Studies in Indian Archaeology, Felicitation Volume*, ed. S.B. Deo, and M.K. Dhavalikar, Bombay: Popular Prakashan, 1985, pp. 75–83.
Matschoss, Conard. *Great Engineers*, London: G. Bells and Sons, 1939.
Maurya, D.M., M.G. Thakkar, A.K. Patidar, S. Bhandari, B. Goyal and L.S. Chamyal. 'Late Quaternary Geomorphic Evolution of the Coastal Zone of Kachchh, Western India', *Journal of Coastal Research*, 2008, vol. 24, pp. 746–758.
McCormick, F. 'Struell: Bathing at Midsummer and the Origins of Holy Wells', in *Religion, Cults and Rituals in the Medieval Rural Environment. Ruralia XI*, ed. C. Bis-Worch, and C. Theune, Leiden: Side Stone Press, 2017, pp. 69–77.
McGovern, Wm Montgomery. 'Notes on Mahayana Buddhism', *The Monist*, 1991, vol. 29, no. 3, pp. 381–382.
Middleton, Nick. *Rivers: A Very Short Introduction*, Oxford: Oxford University Press, 2012.
Mitra, Debala. *Excavations at Tilaura-Kot and Kondon and Explorations in the Nepalese Tarai*, Calcutta: Archaeological Survey of India, 1972.
Mitra, Rajendralala. *The Lalitavistara or Memoirs of the Early Life of the Śākya Sinha*, Calcutta: Asiatic Society, 1882.
Mitra, S.K. 'Ancient Barrage at Pitalkhora', in *Puraratna-emerging Trends in Archaeology, Art, and Anthropology, Conservation, and History, In Honour of Shri Jagat Pati Joshi*, ed. C. Margabandhu, A.K. Sharma, and R.S. Bisht, vol. III, Delhi: Agam Kala Prakashan, 2002, pp. 696–701.
Moore, A.W. and John F. Terry. 'Water and Well-Worship in Man', *Folklore*, vol. 5, no. 3, 1894, pp. 212–229.
Mukhopadhyaya, P.C. *A Report on a Tour of Exploration of the Antiquities in the Tarai, Nepal, the Region of Kapilvastu, during February and March 1899*, Calcutta: Office of the Superintendent of Government Printing, India, 1901.
Murray, G.W. 'Water from Desert: Some ancient Egyptian Achievements', *The Geographical Journal*, vol. 121, 1955, pp. 171–181.
Nāṇdassana, Bhikkhu. *Ratana Sutta, The Buddha's Discourse on the Three Superb Jewels and Their Intrinsic Power- with an Introduction, Translation, and Explanation based on Pali Sources*, Colombo: Sridevi Printers, 2011.
Narada, A Manual of Buddhism, Kuala Lumpur: Buddhist Missionary Society, 1992.
Narayanan, Vasudha. 'Water, Wood, and Wisdom: Ecological Perspectives from Hindu Traditions', *Daedalus*, vol. 30, no. 4, *Religion and Ecology: Can the Climate Change?*, 2001, pp. 179–206.

Nigam, R., R. Dubey, R. Saraswat, A. Sundaresh, S. Gaur and V.J. Loveson. 'Ancient Indians (Harappan Settlement) Were Aware of Tsunami/Storm Protection Measures: A New Interpretation of Thick Walls at Dholavira, Gujarat, India', *Current Science*, vol. 111, no. 12, 2016, pp. 2040–2043. https://doi.org/10.18520/cs/v111/i12/2040-2043

Norphel, Chewang and Padma Tashi. 'Snow Water Harvesting in the Cold Desert in Ladakh: An Introduction to Artificial Glacier', in *Mountain Hazards and Disaster Risk Reduction. Disaster Risk Reduction*, ed. H. Nibanupudi, and R. Shaw, Tokyo: Springer, 2015, pp. 199–210. https://doi.org/10.1007/978-4-431-55242-0_11

Nuttle, W.K. 'Is Ecohydrology One Idea or Many?', *Hydrological Science Journal*, vol. 45, no. 5, 2002, pp. 805–807.

Oldham, C.F. *The Sun and the Serpent, A Contribution to the History of Serpent Worship*, New Delhi: Gyan Publishing House, 2020, reprint.

Oppert, Gustav, *On the Original Inhabitants of Bharatavarsha or India*, Westminster: Archibald Constable and Co., 1894.

Pande, B.M. 'Ring Wells in Ancient India', *Bulletin of the Deccan College Post-Graduate and Research Institute*, vol. 25, Silver Jubilee Volume, 1966, pp. 207–219.

Pandit, Suraj A. 'Water Management System at Kanheri', *Journal of Asiatic Society of Mumbai*, 2010, pp. 27–44.

Panja, Sina. 'A Re-analysis of "Houses": The Evidence from Inamgaon, Maharashtra', *Bulletin of the Deccan College Post-graduate and Research Institute*, vol. 58/59, 1998–99, pp. 93–104.

Park, Chris C. *Sacred Worlds: An Introduction to Geography and Religion*, London: Routledge, 1994.

Pathak, Dinesh and Ananta P. Gajurel. *Climate Change Impacts on Hazards in the Eastern Himalayas*, Kathmandu: International Centre for Integrated Mountain Development, 2010.

Patil, D.R. *The Antiquarian Remains of Bihar*, Patna: K.P. Jayaswal Research Institute, 1963.

Payne, R. *The Canal Builders*, New York: The Macmillan, 1959.

Payutto, P.A. 'Buddhist Economics A Middle Way for the Market Place', 1994, pp. 1–55. https://www.urbandharma.org/pdf/Buddhist_Economics.pdf

Petersen, M.C. *Aggressive Architecture: Fortification of the Indus Valley in the Mature Harappan Phase*, Ph.D. thesis, Leiden University, 2012.

Phillips, W. *Qataban and Sheba: Exploring the Ancient Kingdoms on the Biblical Spice Routes of Arabia*, London: Victor Gollancz Ltd, 1955.

Phuntsho, Karma. *The History of Bhutan*, Noida: Random House, 2016.

Possehl, Gregory L. *The Indus Civilization- A Contemporary Perspective*, New Delhi: Vistar Publications, 2012.

Puri, B.N. 'Irrigation and Agricultural Economy in Ancient India', *Annals of the Bhandarkar Oriental Research Institute*, vol. 48, no. 9, 1968, pp. 383–390.

Queen, Cristopher S. 'Introduction: The Shapes and Sources of Engaged Buddhism', in *Engaged Buddhism; Buddhist Liberation Movements in Asia*, ed. Christopher S. Queen and Sallie B. King, Albany: State University of New York, 1996, pp. 23–35.

Rao, S.R. *Dawn and Devolution of the Indus Civilization*, Aditya Prakashan, New Delhi, 1991.

Rattue, James. *The Living Streams: Holy Wells in Historical Context*, Woodbridge: Boydell Press, 1995.

Reynold, Craig J. 'Power', in *Critical Terms of the Study of Buddhism*, ed. Donald S. Lopez Jr., Chicago: University of Chicago Press, 2005, pp. 211–228.

Reynold, Craig J. 'Magic and Buddhism', in *The Buddhist World*, ed. John Power, London and New York: Routledge, 2016, pp. 338–350.

Reynolds, Frank. 'The Two Wheels of Dhamma: A Study of Early Buddhism', in *The Two Wheels of Dhamma: Essays on the Theravada Tradition in India and Ceylon*,

ed. Gananath Obeyesekere, Frank Reynolds and Bardwell L. Smith, A.A.R. Studies in Religion Series, no. 3 Chambersberg: A.A.R., 1972, pp. 6–31.
Reynolds, Frank. 'Four Modes of Theravada Action', *Journal of Religious Ethics*, vol. 7, Spring, 1979, pp. 12–26.
Reynolds, Frank. 'Multiple Cosmogonies and Ethical Order: The Case of Theravada Buddhism', in *Cosmogony and Ethical Order*, ed. Robin W. Lovin and Frank Reynolds, Chicago: University of Chicago Press, 1985, pp. 203–224.
Rhys Davids, T.W. *Buddhist Birth Stories*, London: Trubner, 1880.
Richard, Gombrich and Gananath Obeysekera. *Buddhism Transformed: Religious Change in Sri Lanka*, Princeton: Princeton University Press, 1988.
Rockhill, W.W. *The Life of the Buddha and the Early History of His Order*, London: Trubner and Company, 2013. reprint.
Ruppert, Brian O. 'Buddhist Rainmaking in Early Japan: The Dragon King and the Ritual Careers of Esoteric Monks', *History of Religions*, vol. 42, no. 2, 2002, pp. 143–174.
Sahni, D.R. 'Sahet-Mahet Grant of Govindachandra [Vikrama] Samvata 1186', *Epigraphia Indica*, vol. XI, pp. 20–26.
Salguero, C. Pierce. 'On Eliminating Disease: Translation of Medical Chapter from the Chinese Versions of Sutra of Golden Light', *Journal of Indian Medicine*, vol. VI, 2013, pp. 21–43.
Samuel, Jeffrey. 'Text Memorized, Text Performed: A Reconstruction of the Role of Paritta in Sri Lankan Monastic Education', *Journal of the International Association of Buddhist Studies*, vol. 28, no. 2, 2005, pp. 339–367.
Sankalia, H.D., Z.D. Ansari and M.K. Dhavilkar. 'Inamgaon: A Chalcolithic Settlement in Western India', *Asian Perspectives*, vol. 14, 1971, pp. 139–146.
Sanyal, N.B. 'Predecessors of the Gahadavālas of Kanauj', *Journal and Proceedings of Asiatic Society of Bengal*, New Series, vol. XXI, 1925, pp. 103–107.
Sarton, G.A. *History of Science: Ancient Science through the Golden Age of Greece*, Massachusetts: Harvard University Press, 1959.
Schmithausen, Lambert. *Problem of Sentiency of Plants in Earliest Buddhism*, Tokyo: The International Institute of Buddhist Studies, 1991.
Schumacher, E.F. 'Buddhist Economics', *Manas*, vol. XXII, no. 33, 1969, pp. 1–14. www.manasjournal.org/pdf
Schweinfurth, Georg A. *Auf unbetretenen Wegen in Aegypten*, Berlin: Hoffman und Campe, 1922.
Sen, Sudipta. *Ganga the Many Parts of a River*, New Delhi: Penguin Random House, 2019.
Sennett, Richard. *Flesh and Stone: The Body and City in Western Civilization*, New York: W.W. Norton, 1994.
Shaw, Julia. 'Archaeologies of Buddhist Propagation in Ancient India: "Ritual" and "Practical models" of Religious Change', *World Archaeology*, vol. 45, 2013a, pp. 83–108, https://doi.org/10.1080/00438243.2013.778132
Shaw, Julia. 'Archaeology of Buddhist Propagation in Ancient India: "ritual" and "practical" Models of Religious Change', *World Arcaheology*, vol. 45, no. 1, 2013b, pp. 83–108, https://doi.org/101080/00438243.2013.778132
Shaw, Julia. 'Naga Sculptures in Sanchi's Archaeological Landscape: Buddhism, Vaishnavism and Local Agricultural Cults in Central India, First Century BCE to Fifth Century CE', *Artibus Asiae*, vol. 64, no. I, pp. 5–59.
Shaw, Julia and John Sutcliffe. 'Ancient dams and Buddhist Landscapes in the Sanchi Area: New Evidence for Irrigation, Land Use, and Monasticism in Central India', *South Asian Studies*, vol. 21, 2005, pp. 1–24.
Shaw, Julia, John Sutcliffe, Lindsay Loyal Smith, Jean-Luik Schwennenger and M.S. Chauhan. 'Ancient Irrigation and Buddhist History in Central India: Optically Stimulated Luminescence Dates and Pollen Sequences from the Sanchi Dams', *Asian Perspectives*, vol. 46, no. 1, 2007, pp. 166–172.

Shraddha, Kumar. *A Self-Governance Approach to Solving the Water Crisis in Ladakh, India: The Ice Stupa Project*, Enschede, Netherlands: University of Twente (Unpublished Master Thesis), 2018–19.
Shukla, S.B., S.P. Prizomwala, V. Ukey, N. Bhatt and L.S. Chamyal. 'Coastal Geomorphology and Tsunami Hazard Scenario along the Kachchh Coast, Western India', *Indian Journal of Geo-Marine Science*, vol. 39, 2010, pp. 549–556.
Shulman, Eviatar. 'The Protective Buddha', *Numen*, vol. 66, no. 2/3, 2019, pp. 207–242.
Singaravelu, S. 'Sītā's Birth and Parentage in the Rāma Story', *Asian Folklore Studies*, vol. 41, no. 2, 1982, pp. 235–236.
Singh, Anand. 'Ecological and Environmental Concerns Amongst the Buddhist Communities of Ladakh: A Case Study', in *Buddhist Virtues and Socio-Economic Development*, MCU Edited Volume, Bangkok: Mahachulalonkorn University, 2011, pp. 428–440.
Singh, Anand. *Planet, Plants, & Animals Ecological Paradigms in Buddhism*, New Delhi: Primus Books, 2019.
Singh, Anand. *Nālandā A Glorious Past*, New Delhi: Primus Books, 2024a.
Singh, Anand. 'Water Sources in the Buddhist Ecology: Looking through the Engaged Tradition', *Journal of Social Innovation and Knowledge*, 2024b, vol. 1, pp. 65–87.
Singh, Anand. 'Giriyaka Hilltop Buddhist Monastic Complex (Rajgir): Understanding Antiquarian Remains and Physical Spaces', *Ancient Asia*, 2021, vol. 12, no. 13, pp. 1–19, https://doi.org/10.5334/aa.251
Sinha, K.K., *Excavation at Sravasti 1959*, Varanasi: BHU, 1959, pp. 5–8.
Sivaramamurti, C. *Amaravati Sculpture in the Madras Government Museum, Bulletin of the Madras Government Museum*, Madras: Thomson and Company Limited, 1956.
Skelton, Andrew. *How Nagas were Pleased by Harsha and The Shattered Thigh by Bhasa*, New York: New York University Press, 2009.
Skilling, Peter. 'The Raksa Literature of the Sravakayāna', *The Journal of Pali Text Society*, vol. XVI, 1992, pp. 109–182.
Skorupski, Tadeusz. *Sarvadurgatiparishodhanatantra, Elimination of All Evil Destiniess*, Sanskrit and Tibetan Text with an Introduction, English Translation and Note, New Delhi: Motilal Banarasidas, 1983.
Spiro, Milford. *Buddhism and Society: A Great Tradition and Its Burmese Vicissitudes*, Berkeley: University of California, 1970.
Stephanie, Kaza and Kenneth Kraft. *Dharma Rain: Source of Buddhist Environmentalism*, Berkely: Shambhala Publication, 2000.
Strang, Veronica. *Water Nature and Culture*, London: Reaktion Books, 2015.
Strong, John S. *The Legend and Cult of Upagupta, Sanskrit Buddhism in North India and Southeast Asia*, Delhi: Motilal Banarasidas, 1994.
Swearer, Donald K. 'Consecrating the Buddha, in *Buddhism in Practice*, ed. Donal S. Lopez Jr., Princeton: Princeton University Press, 1995, pp. 50–58.
Swearer, Donald K. 'An Assessment of Buddhist Eco Philosophy', *Harvard Theological Review*, vol. 99, no. 2, 2006, pp. 123–137.
Swearer, Donald K. *The Buddhist World of Southeast Asia*, Chaing Mai: Silkworm Publishers, 2009.
Swearer, Donald K. *Becoming the Buddha*, New Delhi: Motilal Banarasi das, 2017.
Swinburne, Richard, *The Concept of Miracle*, New York: St. Martin Press, 1970.
Takahashi, Takahiro, Taizo Yamaoka, Fumitako Yoneda, and Akinori Uesugi. 'The Ancient City of Sravasti: Its Significance on the Urbanization of North India', *Puratattva*, no. 30, 1999–2000, pp. 74–92.
Tambiah, S.J. 'The Magical Power of Words', *Man*, vol. 3, no. 12, 1968, pp. 179–208.
Tambiah, S.J. *World Conqueror and World Renouncer, A Study of Buddhism and Polity in Thailand Against a Historical Background*, Cambridge: Cambridge University Press, 1976.

Tambiah, S.J. *Buddhism and the Spirt Cults in Northeast Thailand*, Cambridge: Cambridge University Press, 1979.
Tambiah, S.J. 'The Galactic Polity in Southeast Asia', *HAU: Journal of Ethnographic Theory*, vol. 3, no. 3, 2013, pp. 503–534.
Thaplyal, K.K. *Studies in Ancient Indian Seals*, Lucknow: Akhil Bhartiya Sanskrit Parishad, 1972.
Thera, Piyadassi. *The Book of Protection*, Colombo: H.M. Gunasekara Trust, 1975.
Timmerman, Peter. 'Western Buddhism and Global Crisis', in *Dharma Rain: Sources of Buddhist Environmentalism*. ed. Stephanie Kaza and Kenneth Kraft, Shambhala: Shambhala Publications, 2000, pp. 357–368.
Tinbergen, Jan. 'Wise Management for More Human World', in *Solutions for a Troubled World*, ed. Marc Macy, London: Earth view Press, 1987, pp. 56–57.
Udomittipong, Pipob. 'Thailand's Ecology Monks', in *Dharma Rain: Sources of Buddhist Environmentalism*, ed. Stephanie Kaza and Kenneth Kraft, Boston: Shambala, 2000, pp. 191–197.
Unno, Taitestu. *River of Fire, River of Water: An Introduction to Pure Land Tradition of Buddhism*, London: Double day, 1998.
Vogel, J.P. 'Naga Worship in Ancient Mathura', *ASI Annual Report*, 1908–09, pp. 159–163.
Vogel, J.P. *Indian Serpent-Lore or The Nagas in Hindu Legend and Art*, Delhi: Gyan Publishing House, 2021, (reprint).
Waddel, Laurence. 'The Dharani Cult in Buddhism, Its Origin, Deified Literature and Images', *Ostasiatische Zeitschrift*, 1912, vol. I, pp. 155–195.
Wangchuck, Sonam. 'Ice Stupa Artificial Glaciers of Ladakh', 2014, https://www.indiegogo.com/projects/ice-stupa-artificial-glaciers-of-ladakh
Warder, A.K. *The Path of Discrimination (Patisambhidamagga)*, London: Pali Text Society, 1982.
Westerhoff, Jan. *The Golden Age of Indian Buddhist Philosophy*, Oxford: Oxford University Press, 2018.
Wittfogel, Karl A. *Oriental Despotism: A Comparative Study of Total Power*, New Haven and London: Yale University Press, 1967.
Worster, D. *Rivers of Empire: Water, Aridity, and the Growth of American West*, New York: Pantheon, 1985.
Wulf, H.E. 'The Qanāts of Iran', *Scientific American*, vol. 218, 1968, pp. 94–105.

Reports and Research Volumes

Corpus Inscriptionum Indicarum, Published by Archaeological Survey of India, New Delhi.
Epigraphia Indica, Published by Archaeological Survey of India, New Delhi
Final Report of Assessment of Current Status of Cauvery River in Karnataka and Study of 300 M Buffer Zone on Both Sides of the River, published by Environmental Management and Policy Research Institute (EMPRI), Bengaluru, 2017.
WCD. *The Report of the World Commission on Dams: Dams and Developments-Dams and Development, A New Framework*. London and Sterling: Earthscan Publications Ltd, 2000.

Index

Pages in *italics* refer to figures and pages followed by "n" refer to notes.

FSC
www.fsc.org
MIX
Papier | Fördert gute Waldnutzung
FSC® C083411

Kolibri 360 GmbH
Ferdinand-Jühlke-Straße 7
99095 Erfurt, Deutschland
produktsicherheit@kolibri360.de